高等院校本科生暑期国际设计课程

——绿色社区与低影响开发设计 Workshop

李翅 （加）丹尼尔·罗尔 戈晓宇 周春光 李玥 **编著**

中国建材工业出版社

图书在版编目（CIP）数据

高等院校本科生暑期国际设计课程：绿色社区与低影响开发设计 Workshop / 李翅等编著 . -- 北京：中国建材工业出版社，2020.5

ISBN 978-7-5160-2809-4

Ⅰ. ①高… Ⅱ. ①李… Ⅲ. ①建筑设计－作品集－中国－现代 Ⅳ. ① TU206

中国版本图书馆 CIP 数据核字（2020）第 004930 号

高等院校本科生暑期国际设计课程——绿色社区与低影响开发设计 Workshop
Gaodeng Yuanxiao Benkesheng Shuqi Guoji Sheji Kecheng——Lvse Shequ yu Diyingxiang Kaifa Sheji Workshop
李翅 （加）丹尼尔・罗尔 戈晓宇 周春光 李玥 编著

出版发行：中国建材工业出版社
地　　址：北京市海淀区三里河路 1 号
邮政编码：100044
经　　销：全国各地新华书店
印　　刷：北京天恒嘉业印刷有限公司
开　　本：889mm×1194mm 1/16
印　　张：9
字　　数：200 千字
版　　次：2020 年 5 月第 1 版
印　　次：2020 年 5 月第 1 次
定　　价：99.80 元

本社网址：www.jccbs.com，微信公众号：zgjcgycbs

前言

Preface

绿色社区是以可持续发展思想为指导，作为绿色城市建设的基本单元，实现最高效、最少量、最低影响地使用资源和能源，营造出一种低碳、健康、和谐人性化的聚居环境。相对于传统社区，绿色社区具有较低的环境冲击性、资源和能源消耗性，同时具有良好的人居环境的舒适性以及社会和谐性等特征。

绿色社区的思想可以追溯到英国社会学家霍华德在 1898 年提出的“花园城市”理论，这是绿色社区思想的萌芽。19 世纪 70 年代，麦克哈格所著的《设计结合自然》标志着基于“生态价值观”的社区规划与城市设计成为绿色社区建设的基础。与此同时，欧美国家的绿色社区建设开始进行研究，先后出现了若干应用较广的绿色社区评价体系。

20 世纪末以来，中国许多城市开始思考如何持续发展问题，新的城市发展理念也不断深化。特别是“十九大”提出了关于绿色发展的新理念，国家加快了城市建设领域绿色发展的目标、准则以及指标体系的完善，也相应出台了很多绿色社区建设方面的政策和指导性文件，这些措施的推出标志着中国社区环境建设正向“绿色社区”的目标方向发展。

北京林业大学积极响应绿色社区的发展思想，加强与世界多所著名大学的教学研究合作。在 2018 年 8 月到 9 月，加拿大不列颠哥伦比亚大学建筑与风景园林学院和北京林业大学园林学院举办了为期两周的暑期联合国际课程——《绿色社区与低影响开发设计工作坊》，授课老师为来自加拿大英属哥伦比亚大学（UBC）建筑及风景园林学院的丹尼尔 · 罗尔教授（Daniel Roehr）、助教杰里科 · 班克斯顿（Jericho Bankston），以及北京林业大学园林学院的李翅教授、戈晓宇老师。选课学生为城乡规划、园林、风景园林的本科三年级学生。

丹尼尔 · 罗尔也是温哥华和柏林的注册风景园林师和园艺师，他最重要的工作是在德国柏林的戴姆勒 · 克莱斯勒（Daimler Chrysler）项目波茨坦广场（Potsdamer Platz）水敏性的生活屋顶设计，这项工作是史无前例的开创性的研究项目。本次国际联合课程内容包括以下四个方面：

1. 研究和探讨社区的可持续发展与城市设计策略，以及社区、建筑绿色基础设施的相关内容。

2. 了解低影响开发（LID）的概念以及所有潜在的可能，包括雨水减排过程和生态系统（生态湿地、雨水渗透和管理体系）。

3. 研究和学习如何思考整体建成环境，应用研究成果提出相应的设计策略。

4. 学习如何将当前的 LID 研究应用到一个具体的研究项目地段，如何构建一个 LID 项目并将结果以图纸、图表和论文的形式记录下来，如何通过口头和图形表达技术来展示科学和技术研究数据结果。

本书的内容基于课程研究地段的实践，以北京林业大学校园和龙湖集团高碑店小区项目为例进行探讨低影响开发的绿色社区营建方法与手段，这对于建筑规划、风景园林设计专业人士、环境科学的相关研究人员具有重要的参考价值。本书内容也适合建筑业主、政府管理部门、环境利益相关者和学生阅读。

本书的出版得到北京林业大学创建风景园林学世界一流学科建设引导专项基金资助，同时也得到了北京林业大学教育教学研究重点项目资助，项目名称：规划体系变革下城乡规划专业教育适应性研究（编号：BJFU2019JYZD001）。

李翅

北京林业大学园林学院教授、博士生导师

教育部高等学校城乡规划专业教学指导分委员会委员

中国城市规划学会风景环境规划设计学术委员会副主任委员

目 录

CONTENTS

Advanced Low Impact Development (LID) Workshop BFU, Beijing, China 低影响开发（LID）工作坊课程计划 001

Comprehensive reconstruction of rain and flood management in Beijing Forestry University 北京林业大学校园雨洪管理综合改造 004

1. Site condition analysis 场地条件分析 004

1.1 Site basic information 场地基本信息 004

1.2 Planning policy analysis 规划政策分析 006

1.3 Site climate and precipitation analysis 场地气候与降水分析 006

1.4 Green space and open space analysis 绿地与开放空间分析 008

1.5 Site traffic analysis 场地交通分析 010

1.6 Site status and pavement analysis 场地现状与铺装分析 012

1.7 Site drainage system analysis 场地排水系统分析 013

1.8 Analysis of curreut site problems 场地现状问题分析 017

2. Data calculation 数据计算 018

2.1 Rainwater path analysis 雨水路径分析 018

2.2 Chinese calculation method 中式计算方法 019

2.3 UBC calculation method UBC 计算方法 024

3. Design proposals 设计方案 028

3.1 Design of group one 第一组设计方案 028

3.2 Design proposal of group two 第二组设计方案 052

LID in a New Ecological Community 高碑店某居住区景观雨洪管理综合改造 090

1. Site condition analysis 场地条件分析 090

1.1 Site basic information 场地基本信息 090
1.2 Site current situation analysis 场地现状分析 093
1.3 Calculation based on existing plan 根据已有平面图计算 096
1.4 Pre-design calulation 设计前计算 100
2. LID in Gaobeidian 高碑店 LID 项目学习 109
3. Design Results 设计成果 111
3.1 Design results of group one 第一组设计成果 111
3.2 Design Result of group two 第二组设计成果 122
UBC+BFU "Green Community and Low Impact Development (LID) Frontier Design Workshop" Summer Course Ends Successfully UBC+BFU "绿色社区与低影响开发 (LID) 前沿设计工作坊"暑期课程顺利结束 129

Advanced Low Impact Development (LID) Workshop BFU, Beijing, China
低影响开发（LID）工作坊课程计划

August 27, 2018 – September 7, 2018	2018 年 8 月 27 日—9 月 7 日
Schedule	
Issued **18.8.2018**	2018 年 8 月 18 日版课程计划
Updated	已更新

Instructors		指导教师
Assoc. Prof. Daniel Roehr MBCSLA, CSLA, AKB	**(DR)**	丹尼尔 · 勒尔副教授
Prof. Dr. Chi Li	**(CL)**	李翅教授
Lect. Dr. Xiaoyu Ge	**(XG)**	戈晓宇副教授
Jericho Bankston BArch, MUD	**(JB)**	杰里科 · 班克斯顿 建筑学学士

Required Pre-Requisite Reading 课前必读材料

Roehr, D. and Kong, Y., 2010. “Retro-Greening Suburban Calgary: Application of the Green Factor to a Typical Calgary Residential Site.” Landscape Journal 29(2): 124 - 143.

Roehr, D., Fassman-Beck, E., 2015. “Green Roofs in Integrated Urban Water Systems.” Oxford, England: Routledge: 118 - 135.

Roehr, D. and Kong, Y., 2010. “Runoff Reduction Effects of Green Roofs in Vancouver, BC, Kelowna, BC, and Shanghai, P.R. China.” Canadian Water Resources Journal 35 (1): 53 - 68.

Patrick M. Condon. Seven Rules for Sustainable Communities - Design Strategies for the Post-Carbon World. 9781597268202 GF78.C66 2010 307—dc22.

Publisher: Island Press, Suite 300, 1718 Connecticut Ave., NW, Washington, DC 20009.

Aug. 27

Lecture 1: DR/JB - Stormwater Run Off

Lecture 2: CL/XG - Sustainable Development of Community and Urban Design Strategies

Site 1 Visit - BFU Campus

Assignment 1: Analysis of Site 1

8 月 27 日

讲座 1：勒尔 / 班克斯顿　雨洪径流

讲座 2：李翅 / 戈晓宇　可持续发展的社区与城市设计策略

场地 1 调研　北京林业大学校园

作业 1：场地 1 分析

Aug. 28

LID Precedents lit. research

Lecture 3: DR/JB - Influence of Climate on Reduction of Stormwater Run Off, Crop Coeficient & Future Research North America

Lecture 4: DR - Living (Green) Roofs and Stormwater Management

8 月 28 日

低影响开发案例文献研讨

讲座 3：勒尔 / 班克斯顿　北美地区针对气候对于减少雨水流失、作物系数和未来影响的研究。

讲座 4：勒尔　生态屋顶与雨洪管理

Aug. 29

Site 2 Visit - LID Study Project (Table 1)

(Gaobeidian Project, Longhu Group, Hebei Province)

Survey of Site

8 月 29 日

场地 2 调研　低影响开发项目学习（表 1）

龙湖集团在河北省高碑店的项目

场地调研

Aug. 30

Lecture 5: CL/XG - Project Details, Chinese LID Planning Policies, Rules LID Design in China, Crop Coefficient & Future Research in China with Increasing Dense Urban Conditions

Lecture 6: DR/JB - Stormwater Management Options in View of Increased Climatic Variability, Water Storage Capacities & Specified Flood Zones (ie. Parks & Agricultural Lands) to Minimize Urban Flooding Conditions and Utilization of Urban Conditions for Ecological Strategies

Climate Studies, Interpretation and Mapping of Current Climate Data (Excel &Charting Rainfall Events) to Demonstrate Drastic Increase in Rain Events to Visualize the Increase in Rainfall & Draughts in the Region

8 月 30 日

讲座 5：李翅 / 戈晓宇　中国低影响开发规划政策规则与项目详情和高密度城市背景下，中国低影响开发设计、作物系数和未来研究

讲座 6：勒尔 / 班克斯顿　针对气候变异性、蓄水能力和特定行洪区（即公园与农业用地）增加的雨水管理方案，以最大限度地减少城市洪水并利用城市条件进行生态战略构建

通过气候研究、解释和图示当前的气候数据（降雨的图表）显示降水的急剧增加，以显示该地区降水和干旱情况的增加

Aug. 31

Stormwater Options.

Lecture 7: DR/JB - Green Roof Calculator

8 月 31 日

雨洪管理方法

讲座 7：勒尔 / 班克斯顿　生态屋顶计算器

(GRC) and LID

GRC Exercise and Application of GRC on site

低影响开发生态屋顶计算练习与计算器在场地的应用

Sept. 1

Studio

9 月 1 日

工作坊

Sept. 2

Studio Progress Review with DR, LC, XG, JR

9 月 2 日

勒尔、李翅、戈晓宇、班克斯顿四位导师对工作坊进度做总结

Sept. 3

Lecture 8: DR/JB - Holistic Stormwater Strategies at Site Scale

Lecture 9: LC/XG - Holistic Stormwater Strategies at Urban, Rregional Scale in China

9 月 3 日

讲座 8：勒尔 / 班克斯顿 场地尺度的整体雨洪策略

讲座 9：李翅 / 戈晓宇 中国城市、区域尺度的整体雨洪战略

Sept. 4

Lecture 10: JB - Application of Green Roof Calculator in China

9 月 4 日

讲座 10：班克斯顿 生态屋顶计算器在中国的应用

Sept. 5

Studio Desk Reviews with LC, JB, XG

9 月 5 日

李翅、班克斯顿、戈晓宇工作坊点评

Sept. 6

Studio Desk Reviews with LC, JB, XG

9 月 6 日

李翅、班克斯顿、戈晓宇工作坊点评

Sept. 7

Studio Final Review with LC, JB, XG

9 月 7 日

李翅、班克斯顿、戈晓宇工作坊最终点评

表 1 Site 2 Visit to Gaobeidian Project, Longhu Group, Hebei Province
场地 2 调研：河北省高碑店市龙湖集团项目

Serial number 序号	**Time** 时间	**Scheduling** 日程	**Time control** 时间控制
1	8:00—9:30	*Head to Gao Beidian* 前往高碑店	90 min 90 分钟
2	9:30—11:00	*Visit Gao Beidian Park and Passive Housing Exhibition Hall* 参观高碑店公园和被动房展馆	90 min 90 分钟
3	11:00—12:00	*Visit the construction site* 参观工地	60 min 60 分钟
4	12:00—13:30	*Lunch and break* 午餐及休息	90 min 90 分钟
5	13:30—14:00	*Visit the Aorunshunda office park and the greenhouse* 参观奥润顺达公司与温室	30 min 30 分钟
6	14:00—14:30	*The team of Canadian professors introduce relevant practical cases* 加拿大教授团队介绍相关实际案例	30 min 30 分钟
7	14:30—15:30	*Overall introduction of Gao Beidian project and report relevant design plan* 高碑店项目总体介绍及相关设计方案汇报	60 min 60 分钟
8	15:30—16:30	*Dialogue and communication* 讨论交流	60 min 60 分钟
9	16:30	*Return* 返程	—

Comprehensive reconstruction of rain and flood management in Beijing Forestry University 北京林业大学校园雨洪管理综合改造

Project date: August 2018 项目日期：2018 年 8 月

1. Site condition analysis 场地条件分析

1.1 Site basic information 场地基本信息

图 1 Location of Beijing
北京市区位

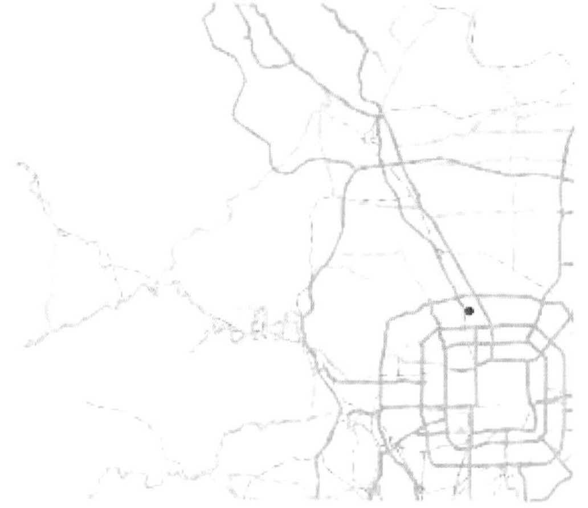

图 2 Location of Haidian district
海淀区区位

图 3 Water system
区域水系

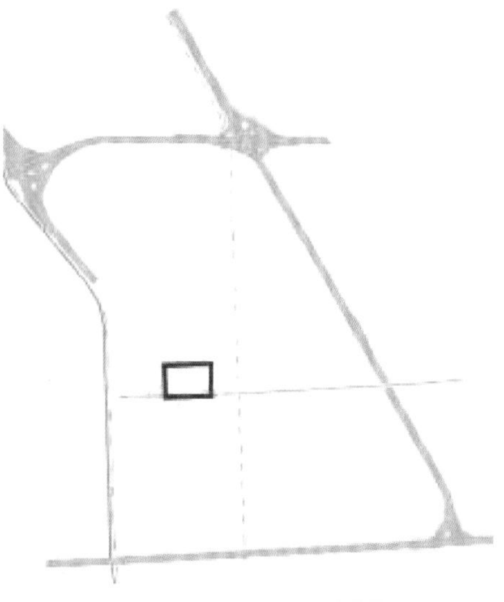

图 4 Location of BFU
北京林业大学区位

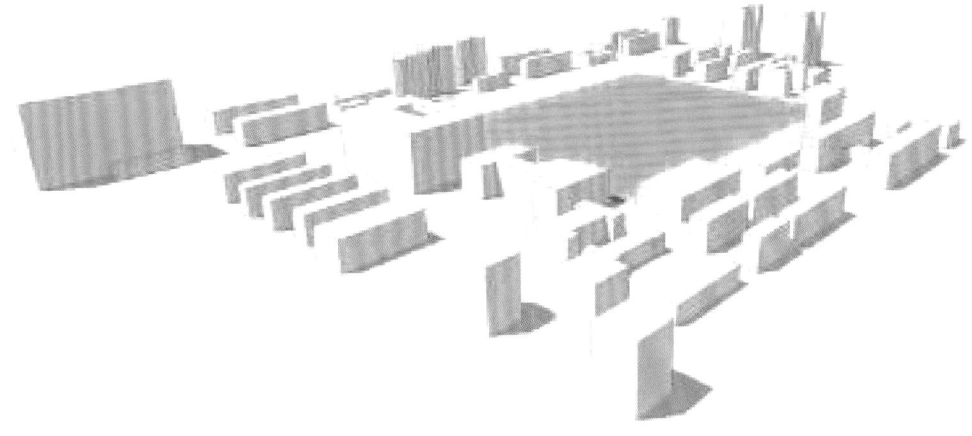

图 5　Building height outside
场地外建筑高度示意图

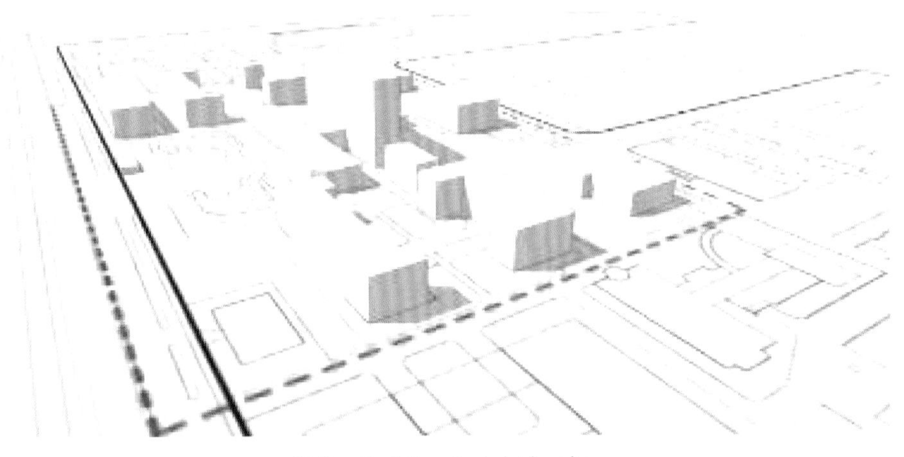

图 6　Building height in site
场地内建筑高度示意图

Beijing Forestry University is located in the east of Haidian District, Beijing. It is adjacent to College Road on the west side and not far from the Fifth Ring Road on the North side. There are two main water systems of Haihe and Ji canal in Beijing, which are rich in water resources. The site is surrounded by Fuhai Lake in the Old Summer Palace, Kunming Lake in the Summer Palace and Qingshui River in Xishan Mountain range. The nearest water body is the Xiaoyue River. It is also the main channel to exclude rainwater from the surrounding area.

北京林业大学位于北京市海淀区东部，西侧毗邻学院路，北侧与五环路相距不远。北京有海河、蓟运河两大主要水系，水资源较为丰富。场地周边有圆明园的福海、颐和园的昆明湖、位于西山山脉的清水河等。距离场地最近的水体为小月河，也是周边雨水汇集排除的主要通道。

The design site is located in the interior of Beijing Forestry University, from the south to the South Gate, from the north to the Baiwax Avenue in front of the library, from the east to the east of the Biological Building and from the west to the east

设计场地位于北京林业大学内部，南至南校门，北至图书馆前洋白蜡大道，东至生物楼东侧道路，西至学研中心东侧道路。周边建筑多为高 20 米左右的板楼，最高为学研中心约 60 米。场地内部建筑多为高 20 米左右的板楼，最高为主楼

of the University Research Center. The surrounding buildings are mostly about 20 meters high, and the highest is about 60 meters. The interior buildings are mostly about 20 meters high, with the highest building about 50 meters.

约 50 米。

1.2 Planning policy analysis 规划政策分析

We will innovate the urban comprehensive management system and promote comprehensive urban management law enforcement. New technologies and concepts, such as sponge city, comprehensive pipe gallery and smart city, are integrated to achieve sound development of urban functions and perfect supporting facilities. We will build urban areas with clean air, clean water, green land and friendly ecological environment, high-standard public transport cities, pedestrian-friendly and bike-friendly urban areas, urban areas with appropriate densities, proper living conditions, balanced working and living conditions, and build national civilized urban areas with clean, beautiful and orderly environment.

(excerpted from Beijing City Master Plan (2016-2035))

创新城市综合管理体制，推进城市管理综合执法。集成应用海绵城市、综合管廊、智慧城市等新技术新理念，实现城市功能良性发展和配套完善。建设空气清新、水清岸绿、生态环境友好的城区，高标准的公交都市，步行和自行车友好的城区，密度适宜、住有所居、职住平衡、宜居宜业的城区，建成环境整洁优美有序的全国文明城区。

（节选自《北京城市总体规划（2016—2035）》）

1.3 Site climate and precipitation analysis 场地气候与降水分析

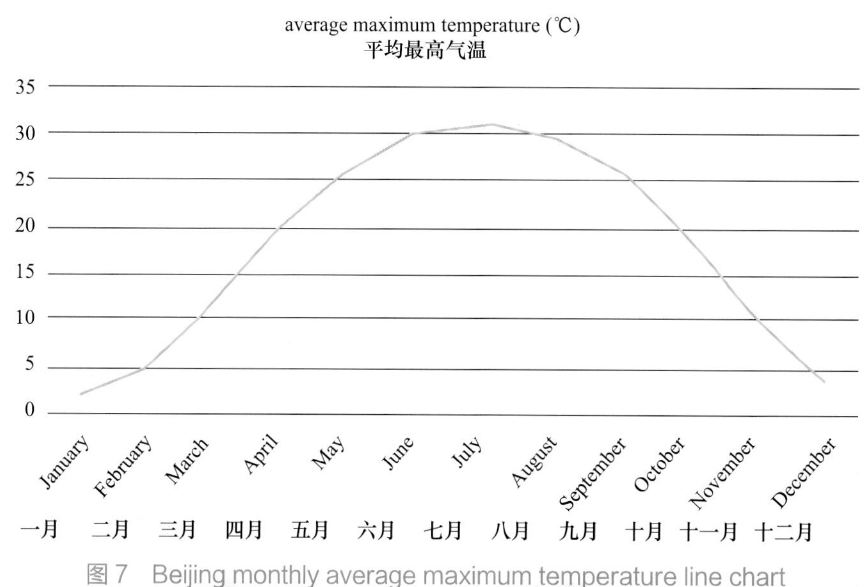

图 7 Beijing monthly average maximum temperature line chart
北京市月平均最高气温折线图

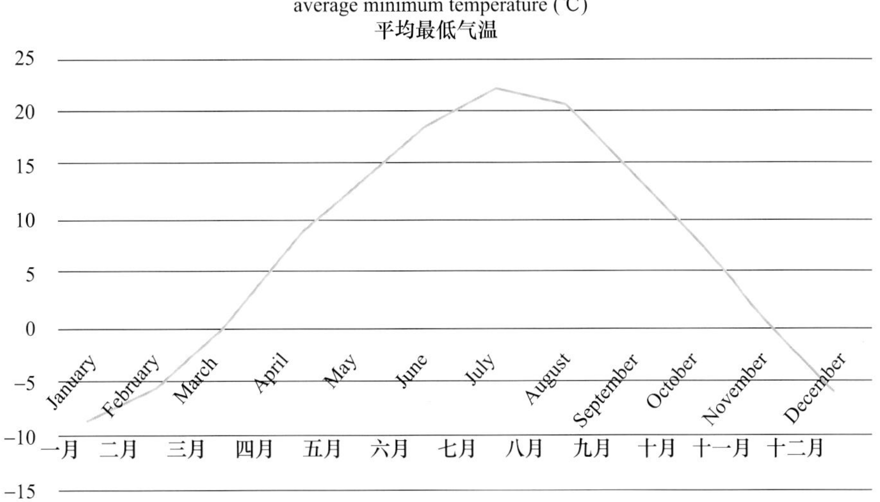

图 8 Beijing monthly average minimum temperature line chart
北京市月平均最低气温折线图

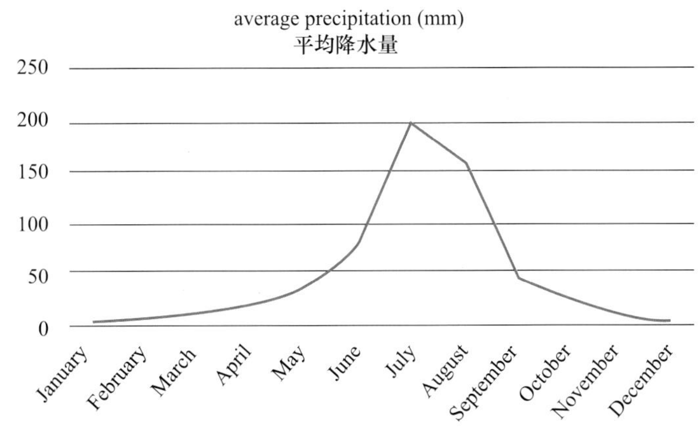

图 9 Beijing monthly average precipitation line chart
北京市月平均降雨量折线图

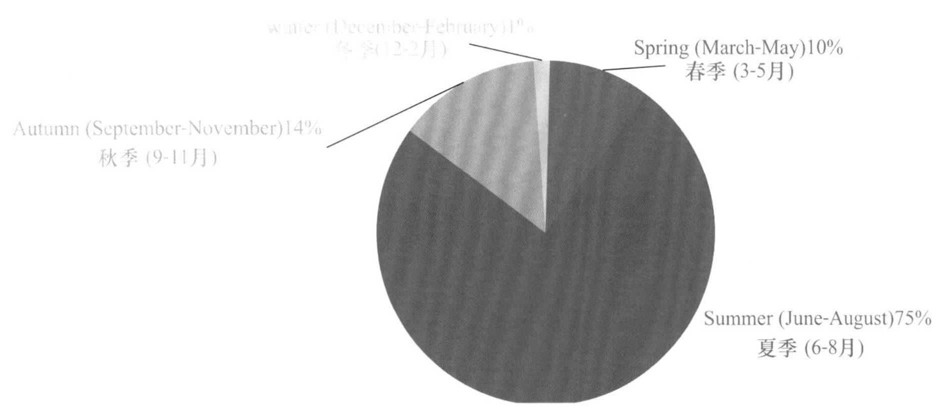

图 10 Haidian seasonal rainfall distribution pie chart
海淀区季度降雨量分布图

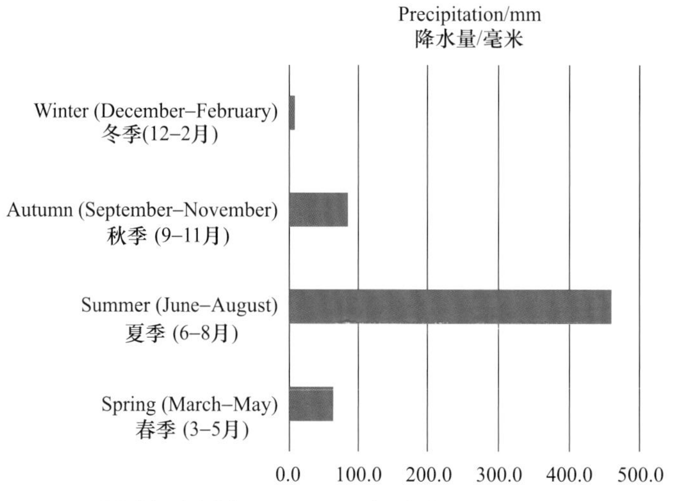

图 11 Haidian seasonal rainfall bar chart
海淀区季度降雨量图

Beijing is located in arid and semi-arid climatic region. The per capita water resources are less than 300mm, which is far below the internationally recognized water shortage limit of 1000mm. Among them, the main source of groundwater recharge is atmospheric precipitation.

北京位于半湿润地区，年均降雨量约600mm，远低于国际公认的1000mm的缺水下限，其中为地下水补给的主要来源是大气降水。

Its landform is: alluvial plain formed by warm humid semi humid monsoon climate.

其地貌类型为：受温带湿润、半湿润季风气候影响所形成的冲积平原。

Haidian District has the largest evaporation in spring and the smallest in winter. Evaporation is greater during daytime than that of night, and plains are bigger than mountains. The annual average evaporation is 1900.4mm. The main evaporation month is from April to June.

海淀区春季蒸发量最大，冬季最小。蒸发量白天大于夜间，平原大于山区。年平均蒸发量为1900.4mm。年内主要的蒸发月份是4—6月。

The precipitation in Haidian District is affected by the monsoon climate and varies greatly from year to year, mostly concentrated in the summer. The average annual precipitation is 614mm, and the spatial and temporal distribution of precipitation is uneven.

海淀区降水受季风气候影响，年际变化较大，主要集中在夏季。年平均降水量为614mm，降水时空分布不均。

1.4 Green space and open space analysis 绿地与开放空间分析

The green space system surrounding Beijing Forestry University is relatively complete, with 10 parks including Baiwangshan Forest Park, the Summer Palace, Yuanmingyuan and Olympic Forest Park. The design site is not well connected to the

北京林业大学周边绿地系统较为完备，共有百望山森林公园、颐和园、圆明园、奥林匹克森林公园等10处公园绿地。而设计场地则与周边的东王庄带状公园、校园绿地连接不佳。设计场地周边存在大量的条带状开敞空间，在设计时应注

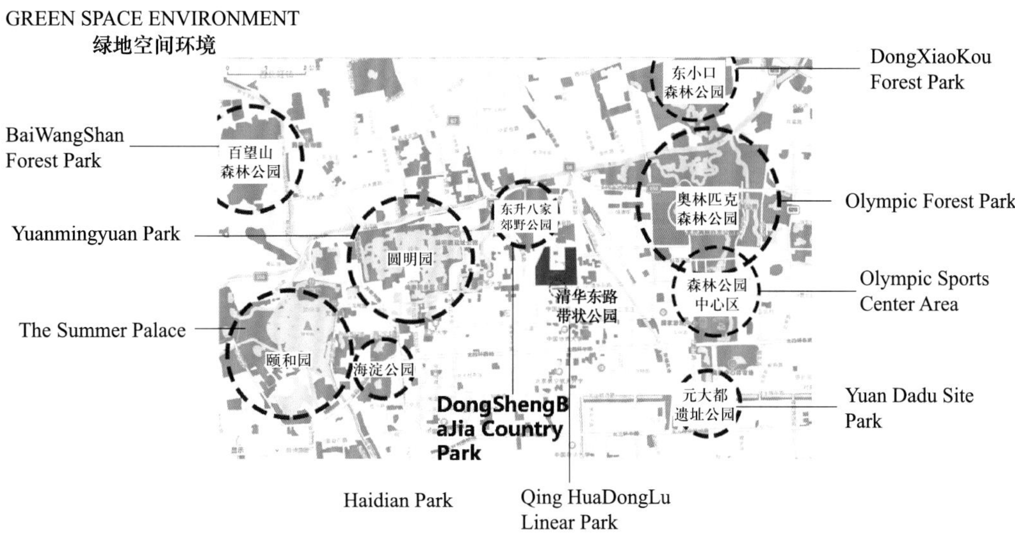

图 12　Regional green space distribution schematic diagram
大区域内绿地分布示意图

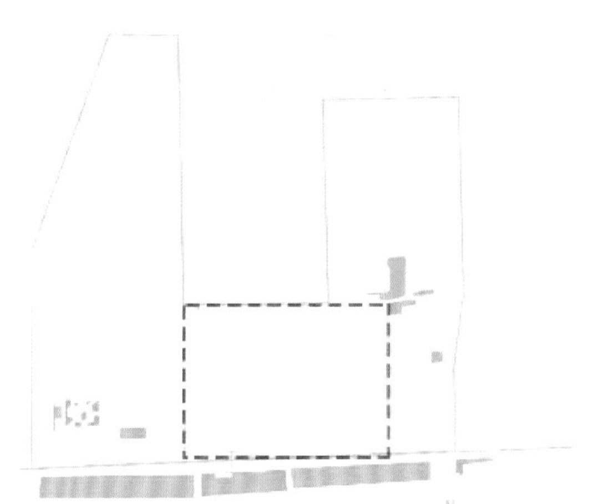

图 13　Surrounding green space distribution schematic diagram
场地周边绿地分布示意图

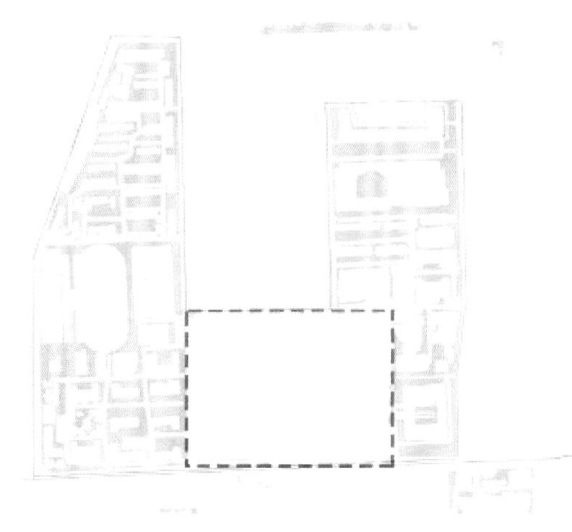

图 14　Surrounding open space distribution schematic diagram
场地周边开敞空间分布示意图

surrounding Dongwangzhuang Ribbon Park and nearby campus green space. There is a large amount of strip-shaped open space around the design site, and attention should be paid to the docking during design.

The campus green space is divided into three categories according to the user's main activities: communicate, take photos and pass by. Through the survey, most of the green spaces are along the roads, and the number of green spaces for viewing and socializing is insufficient.In terms of green space quality, most of the green space is only used

意进行对接。

校园绿地按照使用者的主要活动分为社交、赏景和路过三类。通过调查，校园中绿地的大部分为通过性绿地，用于赏景和社交的绿地数量不足。在绿地质量方面，大部分绿地只是作为楼前绿地，缺乏景观塑造和功能安排。

图 15 Campus green space function diagram
校园绿地使用功能示意图

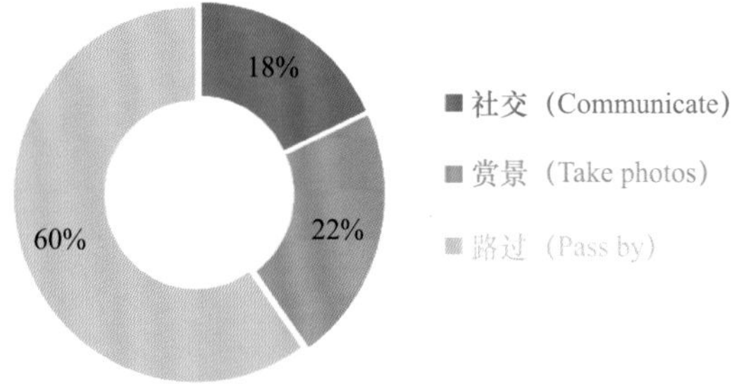

图 16 Campus green space function statistics
校园绿地使用功能统计

TYPE ONE 类别一 利用率高，活动多的绿地

Some greens are useful with many kinds of activities

TYPE TWO 类别二 新建、重建景观良好的绿地

Some greens are recently built or rebuilt with good landscape

TYPE THREE 类别三 楼前绿地

Most greens in our campus are basically surrounding buildings

图 17 Campus green space classification
校园绿地分类

as aspace in front of the building, lacking landscape shaping and functional arrangements.

1.5 Site traffic analysis 场地交通分析

There are 6 open spaces in the site, and the shapes are mostly regular rectangles. There are 6 parking lots in the site, but the number of parking

场地内共有 6 处开敞空间，形状多为较为规整的长方形。场地内现有机动车停车场 6 处，但单个停车场可停车数量不多，从几辆到十几辆不等。

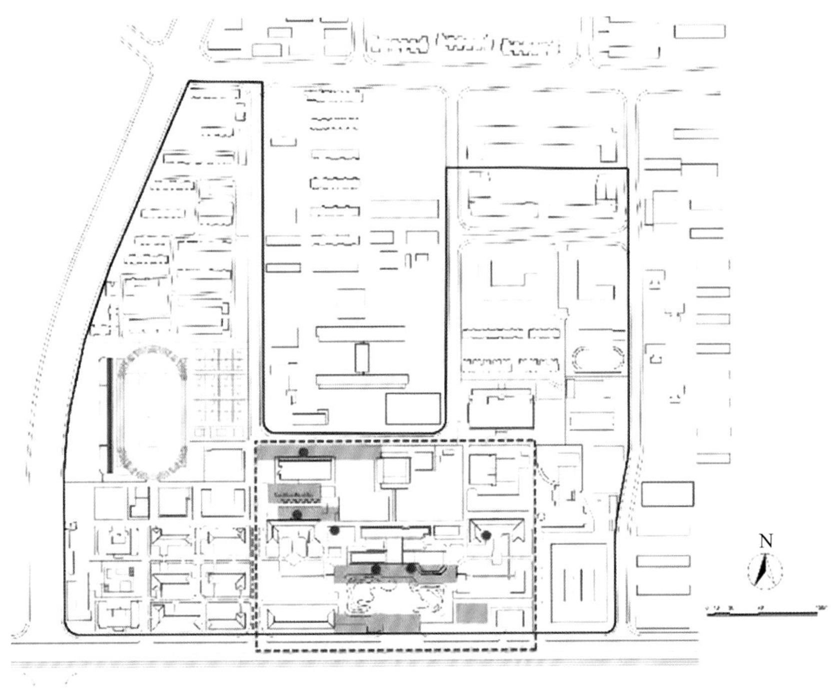

图 18 Open space distribution diagram
开敞空间分布示意图

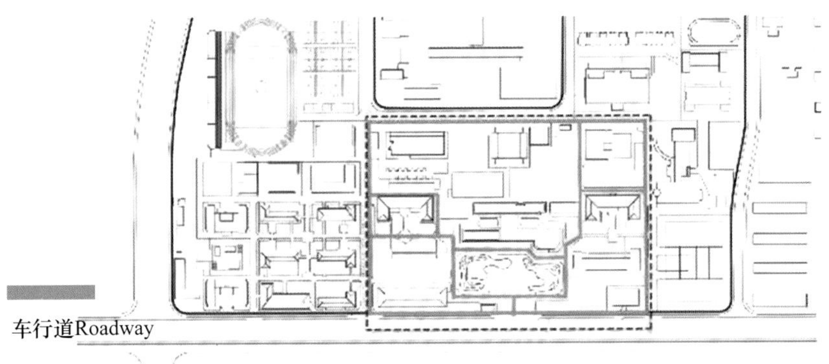

图 19 Driving system analysis
车行系统分析

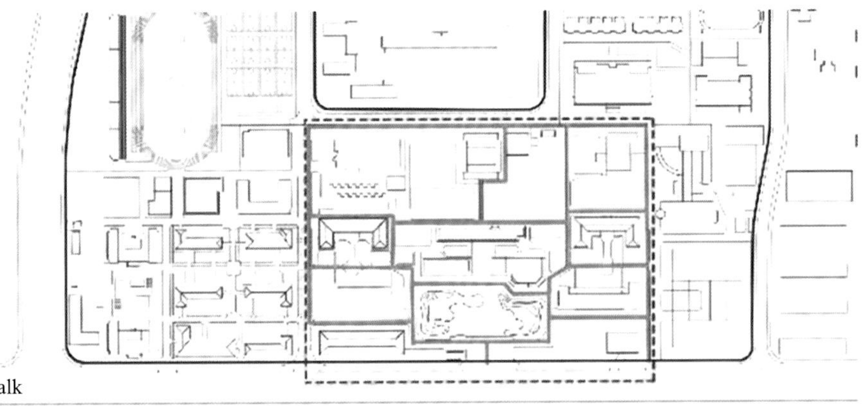

图 20 The pedestrian system analysis
人行系统分析

spaces in each parking lot is small, ranging from a few to more than a dozen.

The accessibility of the roads in the site is relatively high, mostly both for people and vehicles. Except for the sidewalks on the north side, the rest of the roads are shared by pedestrians, non-motor vehicles and motor vehicles. The road on the north side of the main building and on the east side of the school hospital is only for pedestrians and non-motor vehicles.

场地内道路通达性较高，多为人车混行，除北侧道路设有人行道外，其余道路均为行人、非机动车与机动车共享路面。主楼北侧与校医院东侧的道路只供行人与非机动车使用。

1.6 Site status and pavement analysis 场地现状与铺装分析

The current situation of the site is rich in vegetation. The scattered water and rainwater pipes on the building side help to discharge the precipitation to the green space or the flower pond tree pool in front of the building. The rainwater collection pool is designed on the side of some buildings. There are drainage collection devices such as rainwater wells on hard paved roads.

场地现状植被较为丰富，楼侧散水和雨水管有助于将降水排放至楼前绿地或花池树池，部分建筑物侧面设计了雨水收集池。在有硬质铺装的路面上有雨水井等排水收集装置。

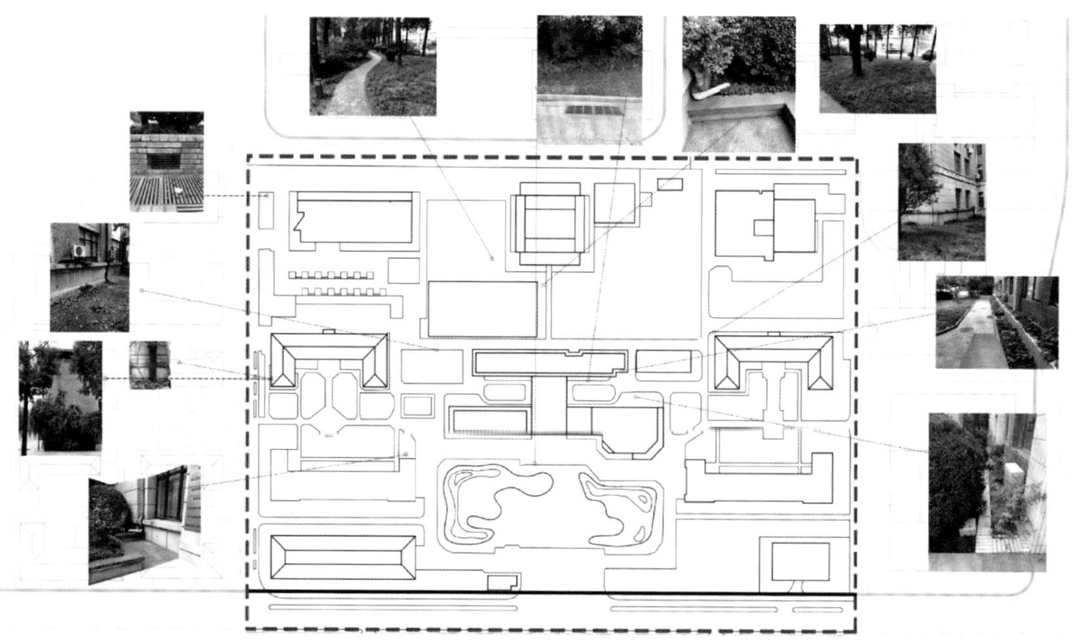

图 21 Site context analysis
场地现状分析

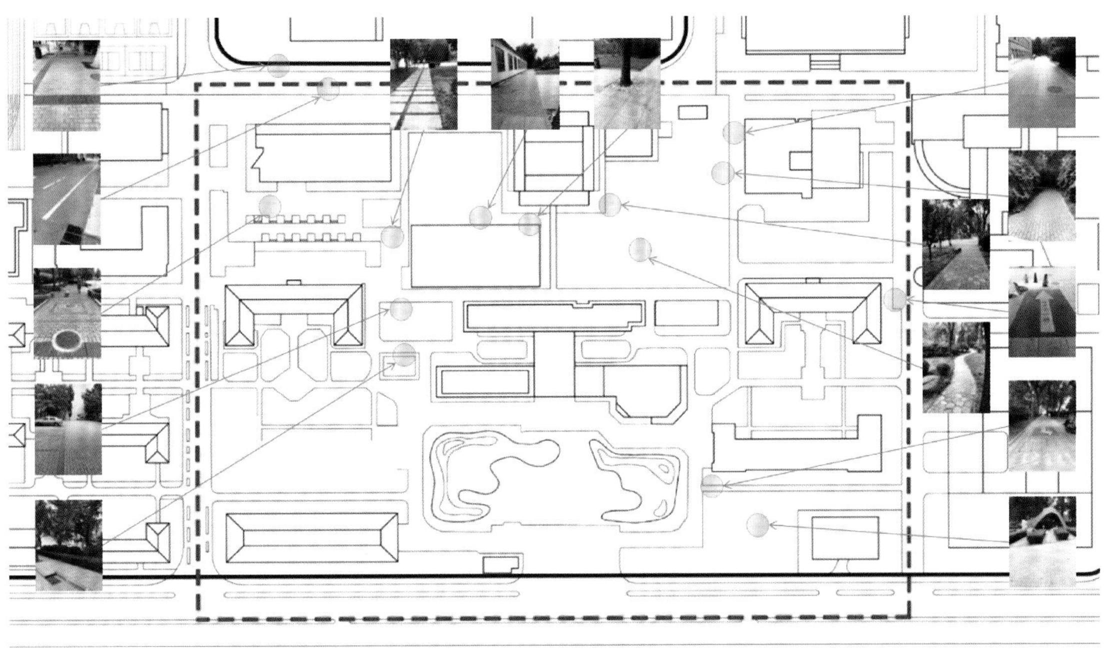

图 22　Pavement analysis
铺装分析

1.7 Site drainage system analysis 场地排水系统分析

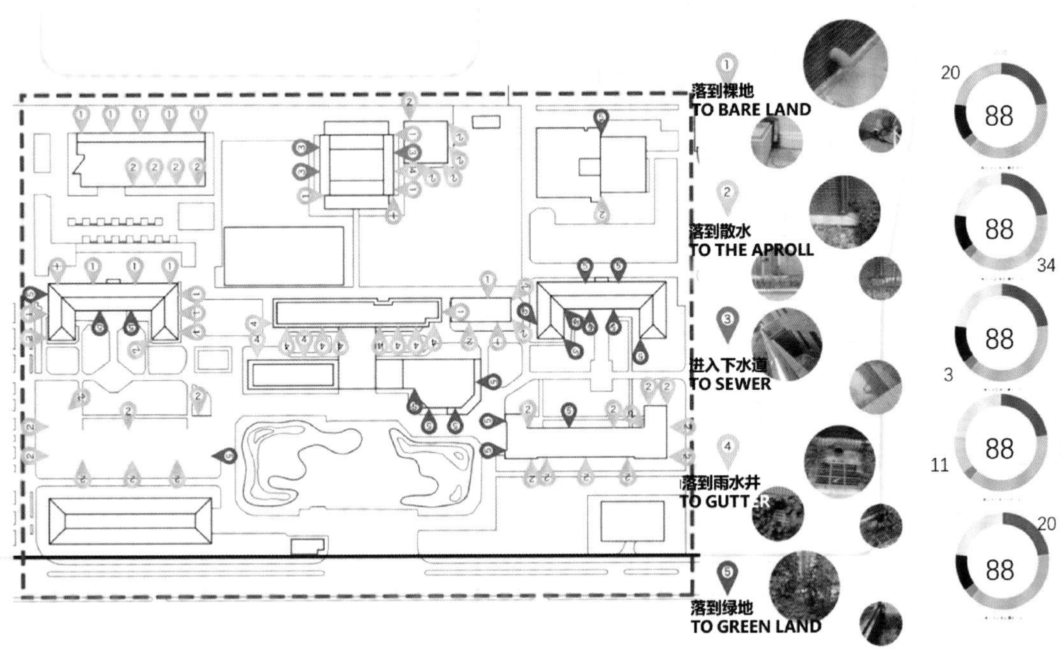

图 23　Building Downpipe Distribution
建筑落水口分布

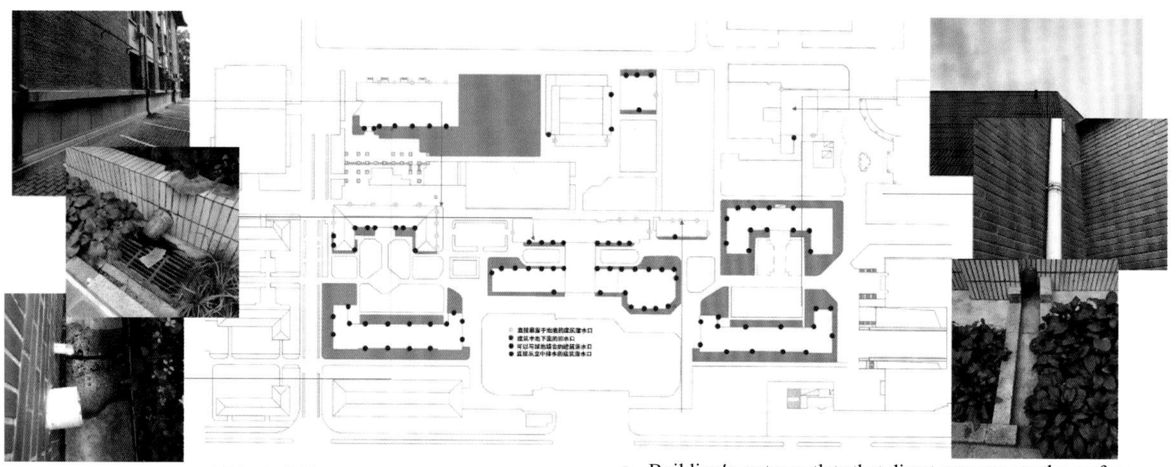

70% Of the building's water outlets can be combined with surrcunding plants, but most need to be landscaped
70%的建筑落水口可结合周边植被设计，但大多数需要美化

- Building's water outlets that direct exposure to the surface
  直接暴露在外的建筑落水口
- Building's water outlets in the basement
  地下建筑排水口
- Building's water outlets that could be designed by combining with Green space
  可结合绿地设计的建筑落水口
- Building's water outlets on the second floor
  位于二层的建筑落水口

图 24　Catch Basin Distribution
场地雨水口分布

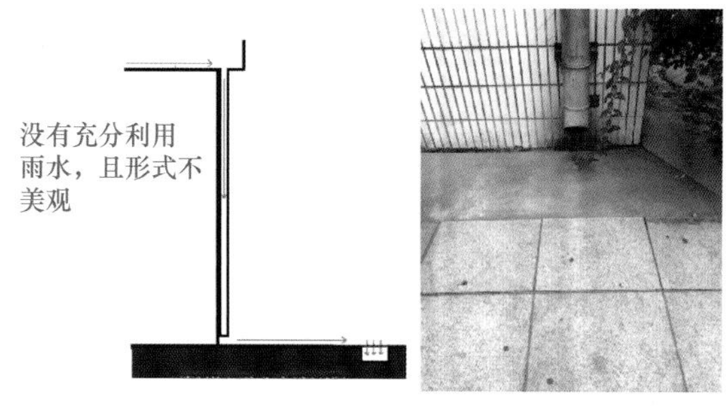

图 25　Directly discharge to the ground
直接排放至地面
Disadvantages: Do not fully utilize rain and the form is not artistic enough.
缺点：没有充分利用雨水且形式不美观

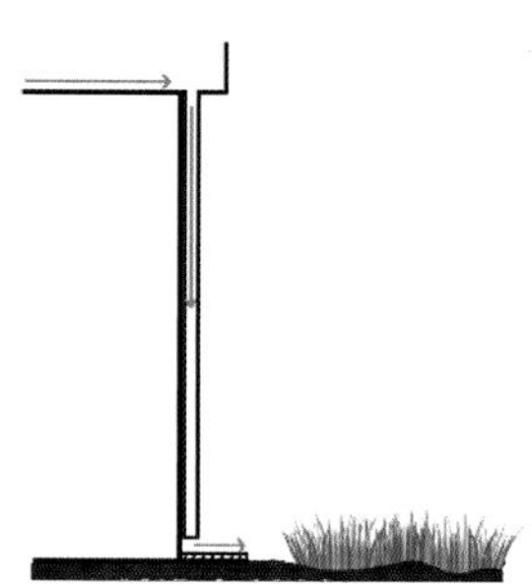

图 26　Indirectly discharge to green space
间接排放至绿地
Disadvantages: none
缺点：暂无

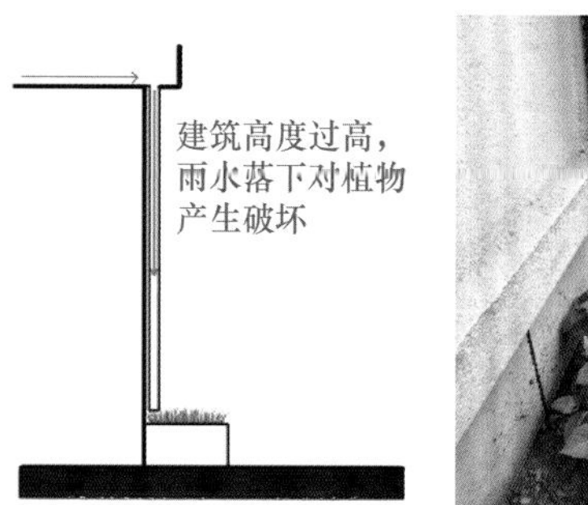

图 27　Directly discharge to green space
直接排放至绿地
Disadvantages: The building is too high, and the rain falls to damage the plants.
缺点：建筑高度过高，雨水落下对植物产生破坏

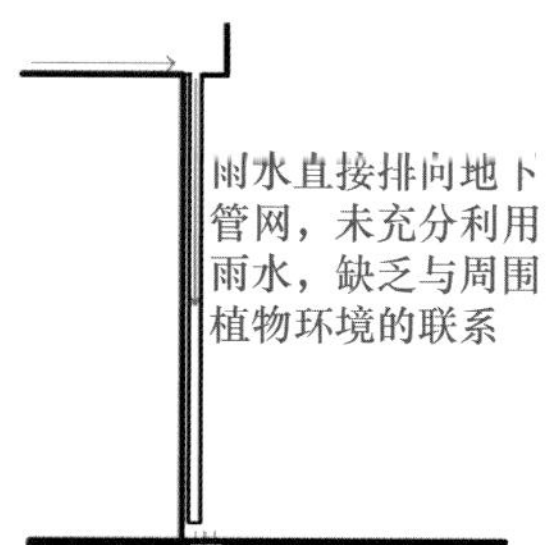

图 28　Directly to the underground pipe network
直接排向地下管网
Disadvantages: Do not utilize the most of rainwater and lack of links to the surrounding plant environment.
缺点：未充分利用雨水，且缺乏与周围植物环境的联系

图 29 使用雨水管排水
Using rainwater pipe to drain
Disadvantages: Do not link to the surrounding plant environment, and do not make full use of rainwater.
缺点：没有和周围植物环境联系，未充分利用雨水

Investigate the drainage mode of the building and the location of the drainage port of the site, and summarize the five drainage modes.

对场地内建筑排水方式与场地的排水口位置进行调查，并总结出 5 种排水模式。

Problems:

1. The impact of rain on the ground and plants.

2. Not making full use of the rain water, cut off from the surrounding plant environment.

3. The building's downspout is in disrepair.

4. Without the water storage device, the overflow increases the drainage pressure of the pipe network.

问题发现：

1. 楼层过高，雨水冲击力破坏地面及植物。

2. 雨水直接排入雨水口中，没有充分利用，与周围植物环境割裂。

3. 建筑落水管年久失修，不美观。

4. 没有储水装置，降雨较大时，溢流到地面，加重管网排水压力。

Strategies:

1. Cushion the end of building downpipes.

2. Surround the green space around the building so that rain water is directly discharged into the green space.

3. Use vegetation to hide it.

4. Set up a small rain garden around the downspout for water storage.

解决策略：

1. 对建筑落水管末端设置缓冲装置。

2. 在建筑周围围合绿地，使雨水直接排入绿地中。

3. 对落水管进行修缮，可将其隐藏在建筑内部，也可以利用植被将其隐藏。

4. 在落水管周围设置小型雨水花园，用于储水。

There are 4 types of tree pools in the site, which are tree pools with 4-5cm high guardrails, tree pools with soil below the road surface and covered by pebbles, tree pools with bare soil below the road surface , and tree pools set up with grass.

场地内现有 4 种类型的树池，分别为有 4 ~ 5 厘米高的护栏的树池，土壤低于路面并被卵石覆盖的树池，土壤低于路面并被土壤裸露的树池，以及和草地共同设置的树池。

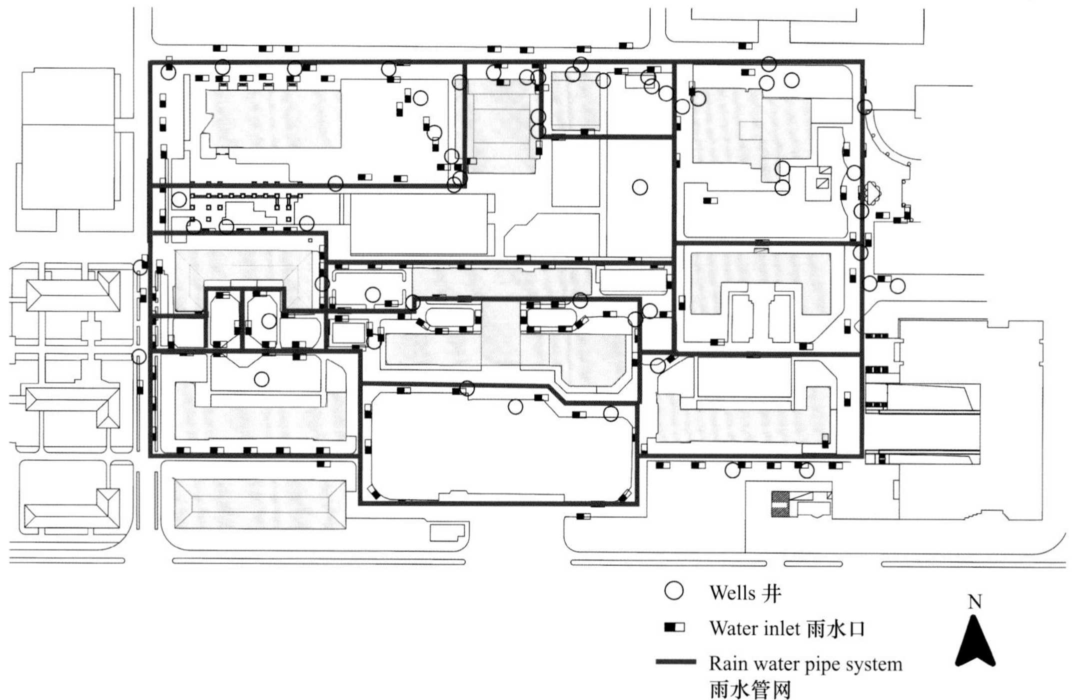

图 30　Current rainwater pipe network
现状场地雨水排水管网

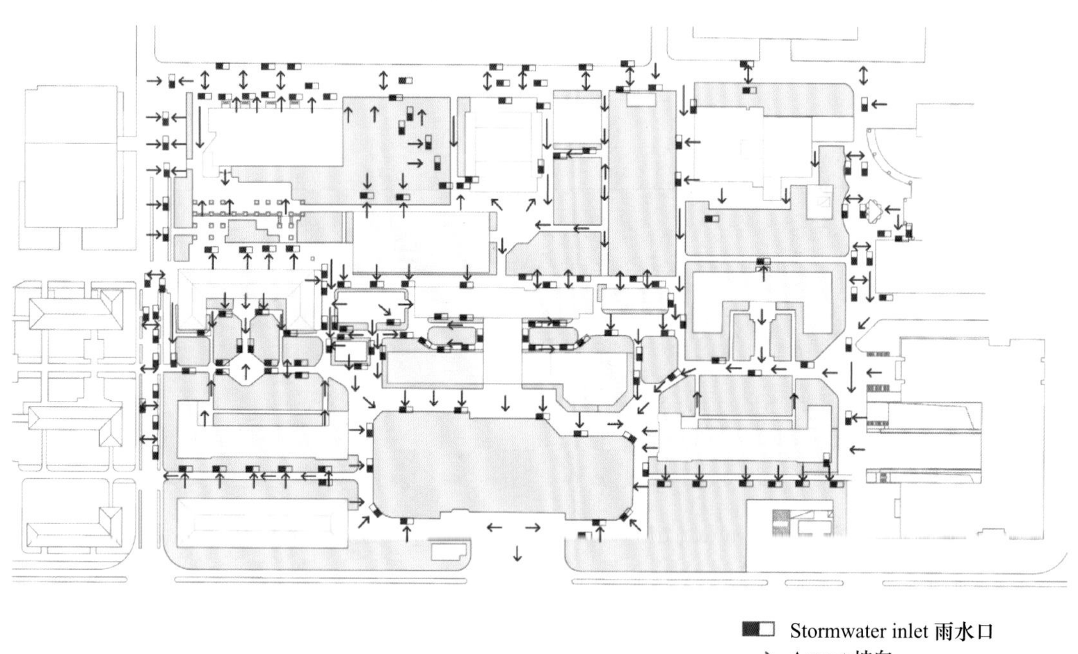

图 31　Current slope of the site
现状场地坡向

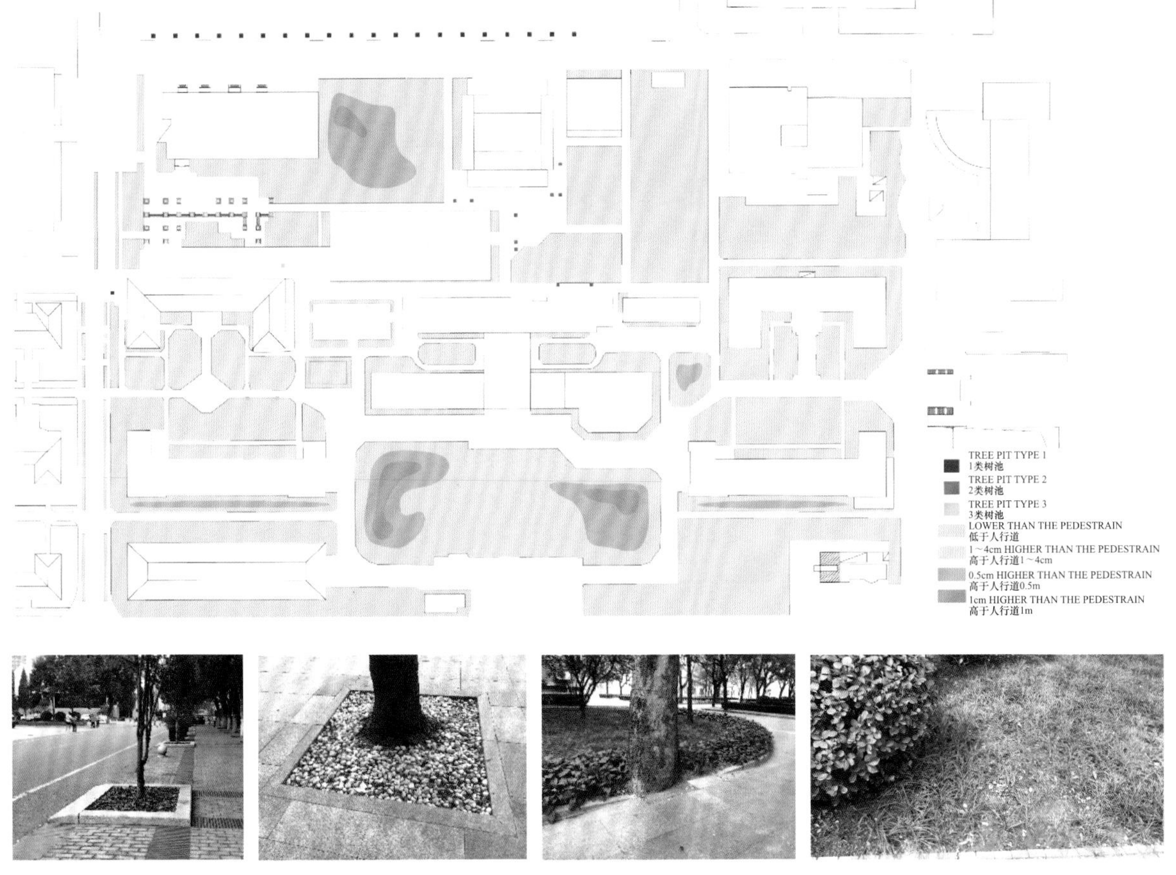

图 32 Regional green space distribution schematic diagram
绿地高程分析

1.8 Analysis of curreut site problems 场地现状问题分析

Erosion—The rain falling from high roofs is strong, but the design for responding to rainwater is simple.

Appearance—Random plant species , shabby building's downspout and the basic green lands are not so beautiful.

Separation—The green space boundary is surrounded by the teeth of roads, which isolates the connection of green space and road surface rain.

Function—Most of the green lands only serve for our eyes, without the ability of stormwater management or serving as temporarily activity spaces.

Irrationality—The slopes of many green lands and the roads are not agreeable to the flow of runoff.

侵蚀——从高屋顶落下的雨水力量强，但雨水的应对设计很简单。

外观——植物种类随意，破旧的建筑落水管和基础绿地缺乏美观性。

分离——绿色空间边界被路缘包围，隔离了绿地和路面雨水的连接。

功能——大多数绿地只服务于视觉，没有雨水管理能力或作为临时活动空间的能力。

非理性——许多绿地和道路的坡向与径流的流向不一致。

2. Data calculation 数据计算

2.1 Rainwater path analysis 雨水路径分析

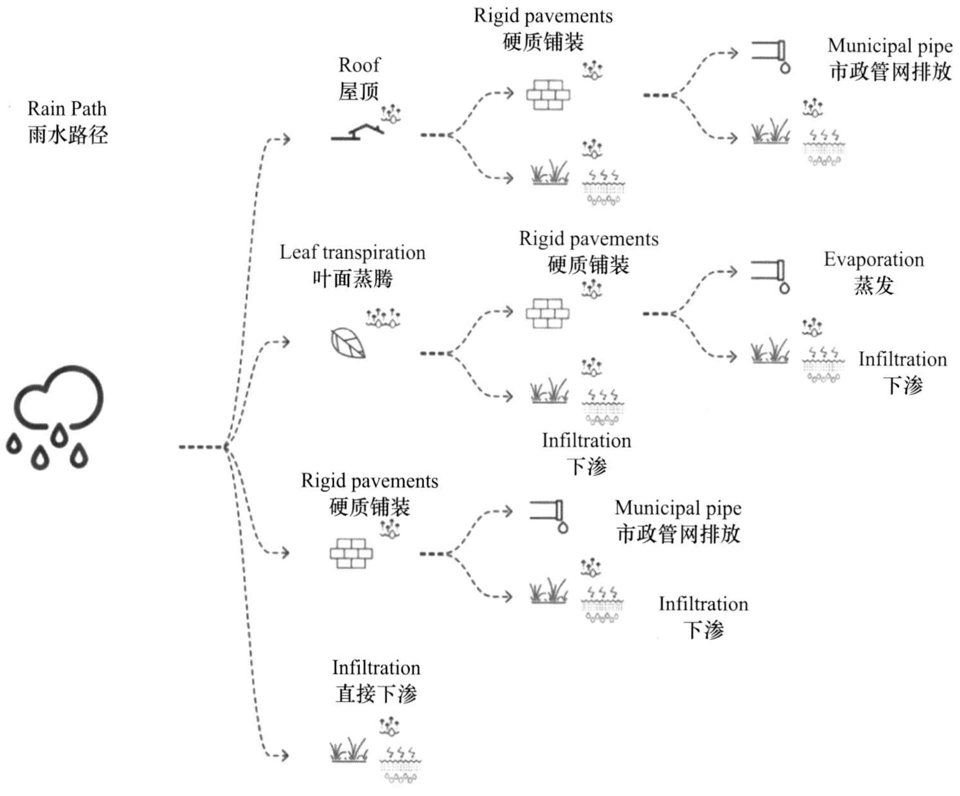

图 33 Rain path schematic diagram
雨水路径示意

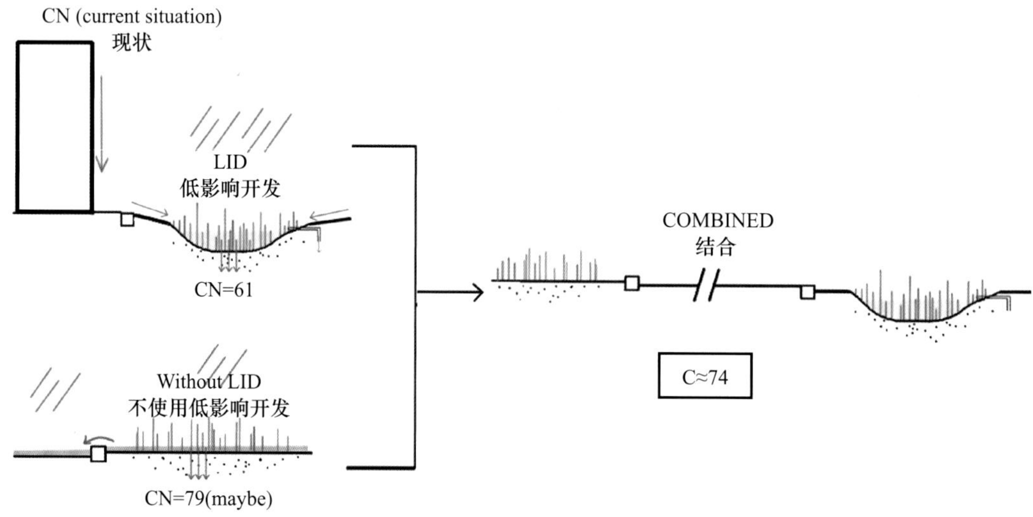

图 34 Technical route of LID
低影响开发技术路线示意

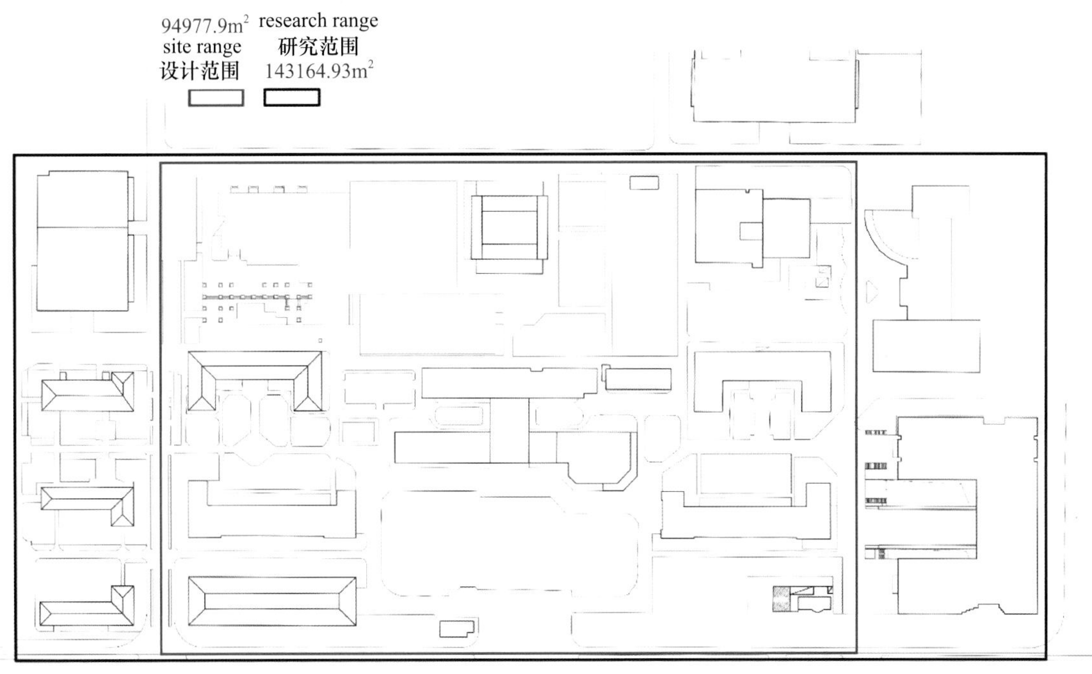

图 34　The scope of the study
研究范围示意

2.2 Chinese calculation method 中式计算方法

$$W=10\psi_{ZC}h_yF$$

W——The total runoff (m^2) 总径流量；

ψ_{ZC}——Rainfall comprehensive runoff coefficient 综合径流系数；

h_y——Designed rainfall (mm) 设计降雨量；

F——Catchment area (hm^2) 汇水面积。

2.2.1 Site Range 设计范围

表 2　Statistics of suface types and area within the site range
场地表面类型与面积统计设计范围

The type of surface 表面类型	**Area (m^2)** 面积	**Percentage** (%) 百分比
green roof 生态屋顶	814.61	0.86%
pavement (impervious) 铺装（不透水）	38599.95	40.64%
green land 绿地	30268.38	31.87%
grey roof 硬质屋顶	21392.56	22.52%
basement which is covered with green land 绿化覆盖的地下室	3902.40	4.11%
sum 总和	94977.90	100.00%

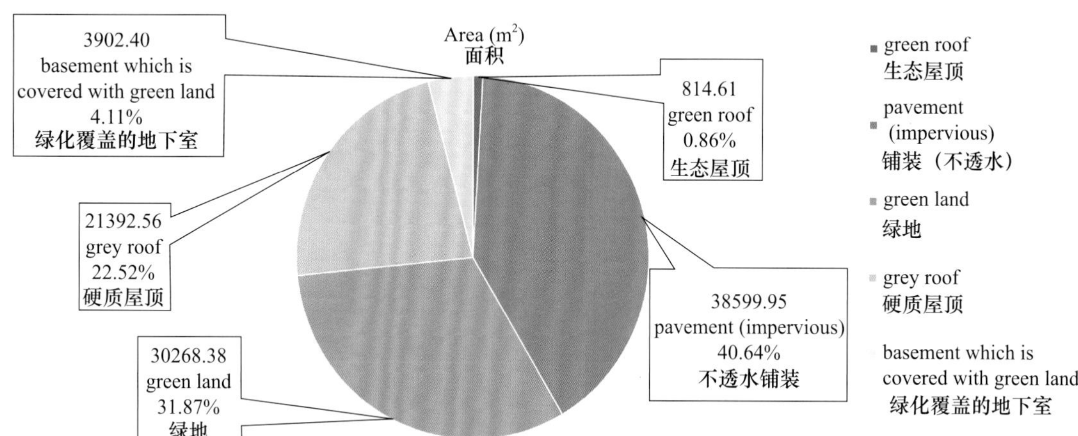

图 35　pie chart of surface types and area within the site range
设计范围场地表面类型与面积饼状图

site range
设计范围
research range
研究范围
38599.95m²
40.64%
pavement
铺装
4717.01m²
4.97%
green roof
(including basement green)
生态屋顶（包含地下室绿化）
grey roof 硬质屋顶
21392.56m²
22.52%
green land 绿地
30268.38m²
31.87%

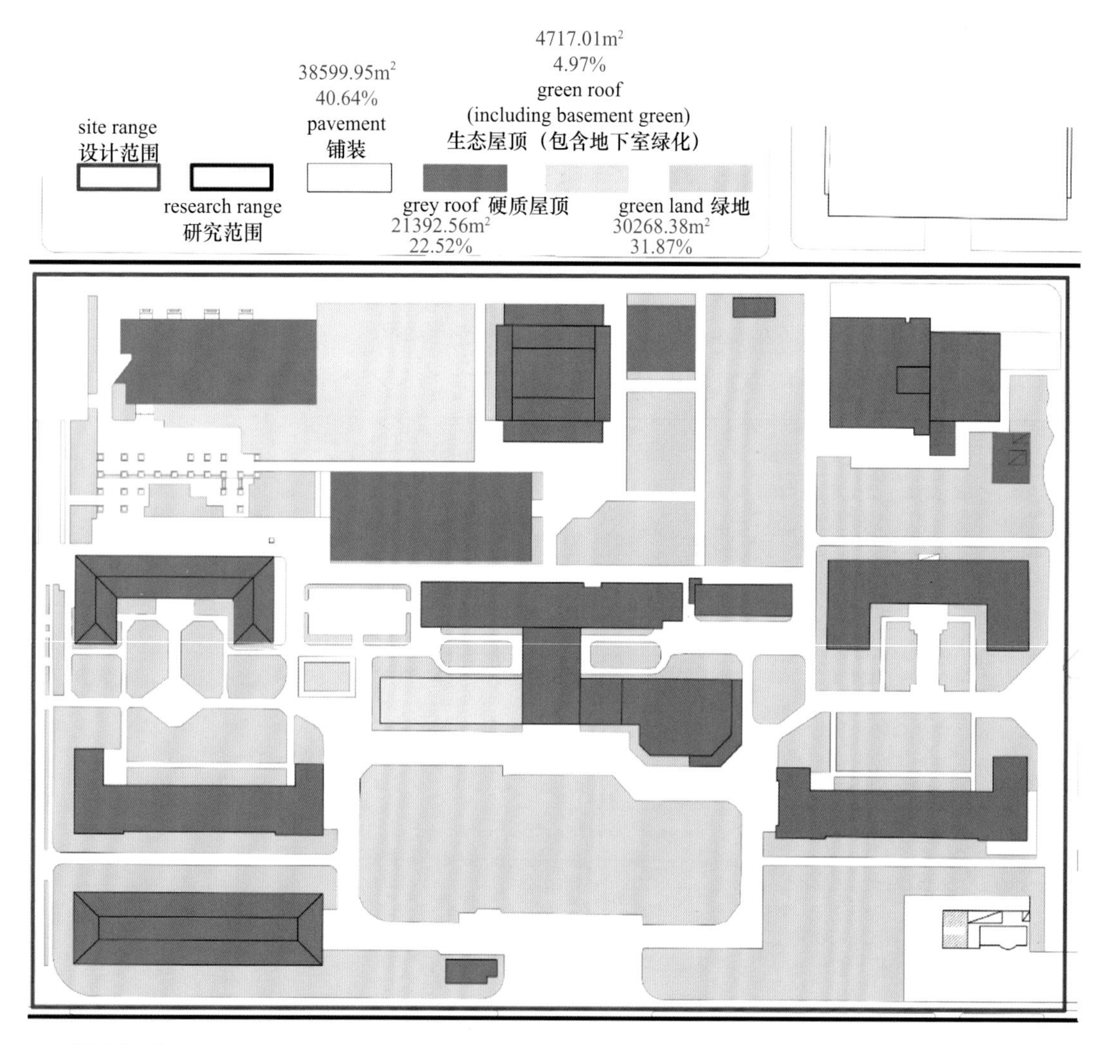

图 36　Schermatic diagram of distribution of different surface types within the site range
设计范围内不同表面类型分布示意图

表 3 Calculation of runoff within the site range
设计范围场地径流量计算

current (site range) 现状（设计范围）	Area (m^2) 面积	Runoff coefficient 径流系数	W (runoff/m^3) 径流量	
Type 类别			10 year period（10 年期）	3 year period（3 年期）
green roof 生态屋顶	814.61	0.30	51.08	26.39
pavement (impervious) 不透水铺装	38599.95	0.90	7260.65	3751.92
green land 绿地	30268.38	0.15	948.91	490.35
grey roof 硬质屋顶	21392.56	0.90	4023.94	2079.36
basement which is covered with green land 绿化覆盖的地下室	3902.40	0.15	122.34	63.22
sum 总和	94977.90	—	12406.92	6411.24

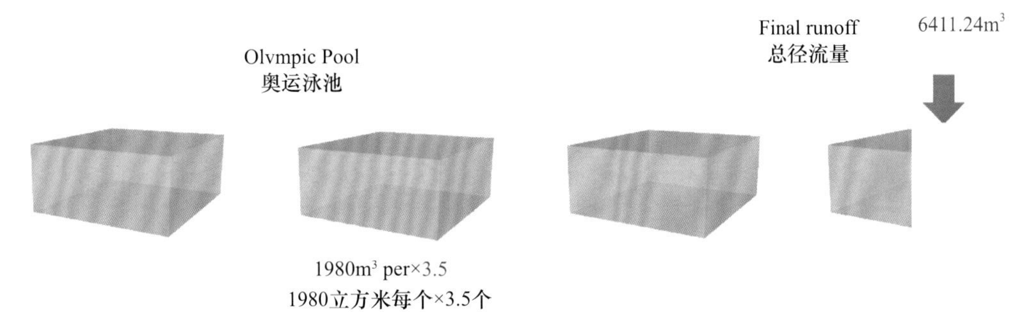

图 37 Schematic diagram of frual runoff within the site range
设计范围场地总径流量示意

2.2.2 Research Range 研究范围

表 4 Statistics of surface types and area within the research range
研究范围场地表面类型与面积统计

The type of surface 表面类型	Area (m^2) 面积	Percentage (%) 比例
green roof 生态屋顶	814.61	0.57%
pavement (impervious) 不透水铺装	65810.85	45.97%
green land 绿地	35817.06	25.02%
grey roof 硬质屋顶	36820.01	25.72%
basement which is covered with green land 绿化覆盖的地下室	3902.40	2.72%
sum 总和	143164.93	100.00%

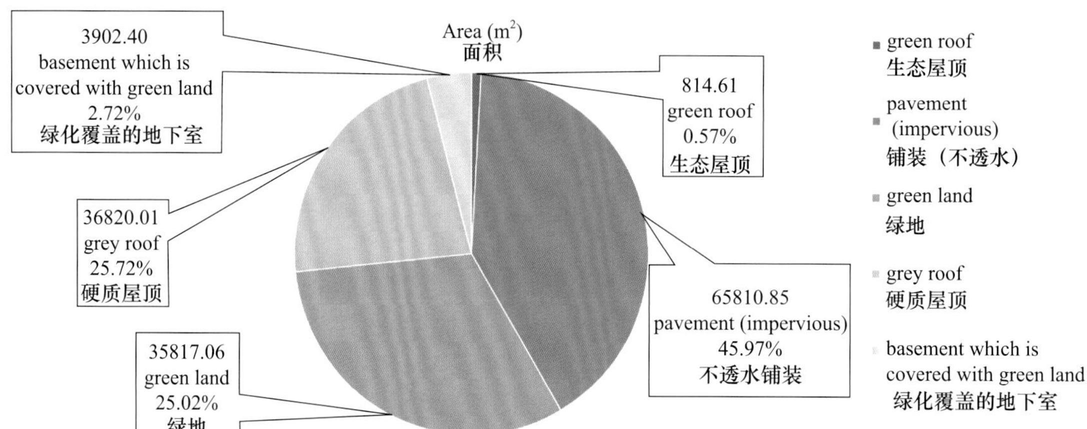

图 38　Pie chart of surface types and area within the research range
研究范围场地表面类型与面积饼状图

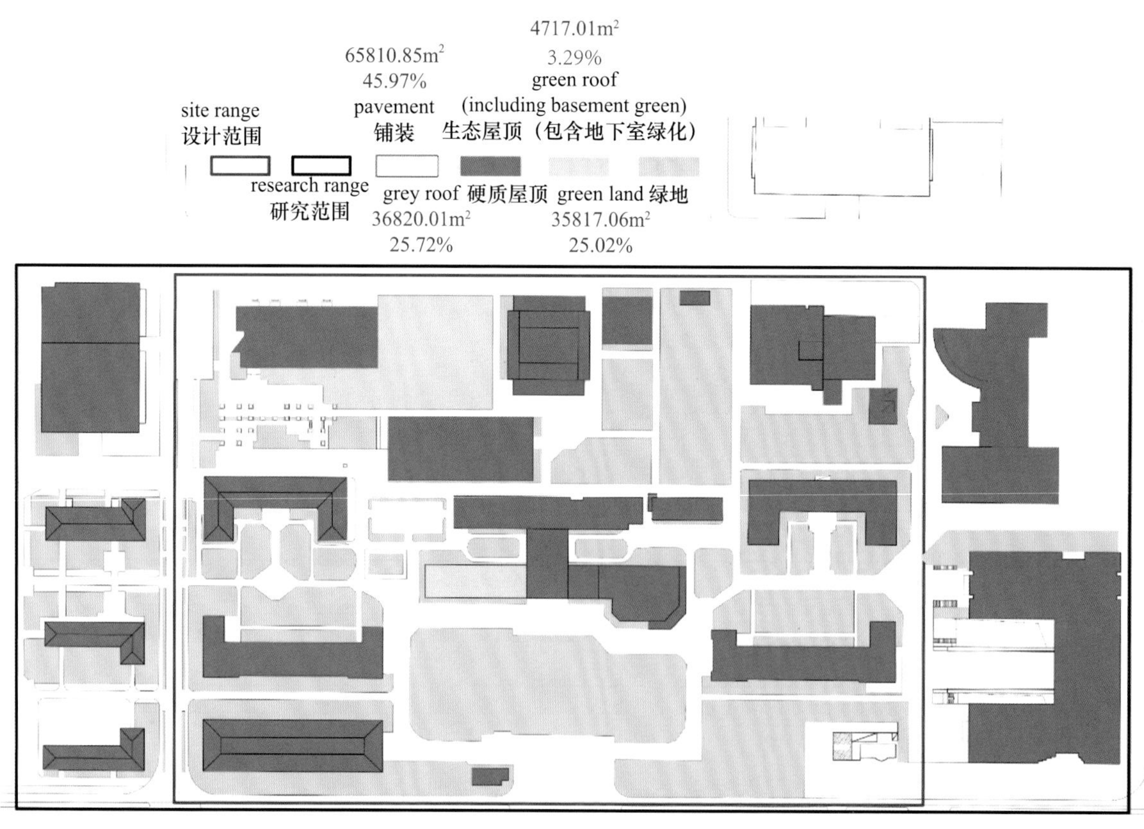

图 39　Schematic diagram of distribution of different swface types within the research range
研究范围内不同表面类型分布示意图

表 5　Calculation of runoff within the reseur range
研究范围场地径流量计算

current (research range) 现状（研究范围）	Area (m^2) 面积	Runoff coefficient 径流系数	W (runoff/m^3) 总径流量	
Type 类别			10 year Event 十年期	3 year Event 三年期
green roof 生态屋顶	814.61	0.30	51.08	26.39
pavement (impervious) 不透水铺装	65810.85	0.90	12379.02	6396.81
green land 绿地	35817.06	0.15	1122.86	580.24
grey roof 硬质屋顶	36820.01	0.90	6925.84	3578.90
basement which is covered with green land 绿化覆盖的地下室	3902.40	0.15	122.34	63.22
sum 总和	143164.93	—	20601.14	10645.56

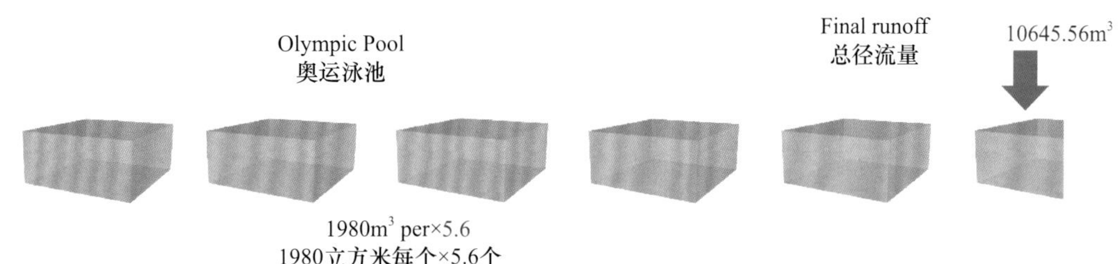

图 40　Schematic diagram of final rwunff wifuin the research range
研究范围场地总径流量示意

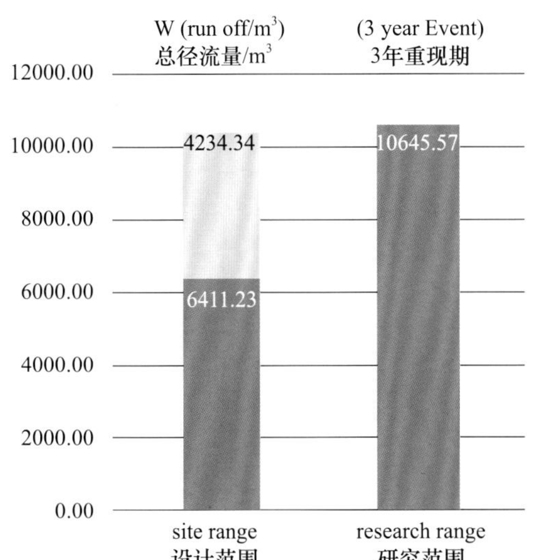

图 41　Comparison of finall runoff and three-year flood data insite range and research range
设计范围与研究范围总径流量与三年期洪水数据比较

Considering the buildings and pavement around our site, there will be another 4234.34m^3 water which may flow to our site and need to be solved in our site. (as the worst situation)

考虑到我们场地周围的建筑物和路面，还有另外 4234.34m^3 的水可能流入我们的场地，需要在我们的场地解决。（最糟糕的情况）

2.3 UBC calculation method UBC 计算方法

2.3.1 Site Range 设计范围

表 6 Statistics of surface types and area within the site range
设计范围场地表面类型与面积统计

	The type of surface 表面类型	Site range/m² 场地面积	Percentage/% 百分比
Roof 屋顶	Roof Area 屋顶面积	① 26109.57	—
	Extensive Rf 粗放型屋顶	4717.01	4.97%
	Intensive Rf 集约型屋顶	0.00	—
	Grey Roof 硬质屋顶	21392.56	22.52%
Ground 地面	Ground Area 地面面积	② 68868.33	—
	Green 绿地	30268.38	31.87%
	Grey 硬质铺装	38599.95	40.64%
	Sum 总和	94977.90	100.00%

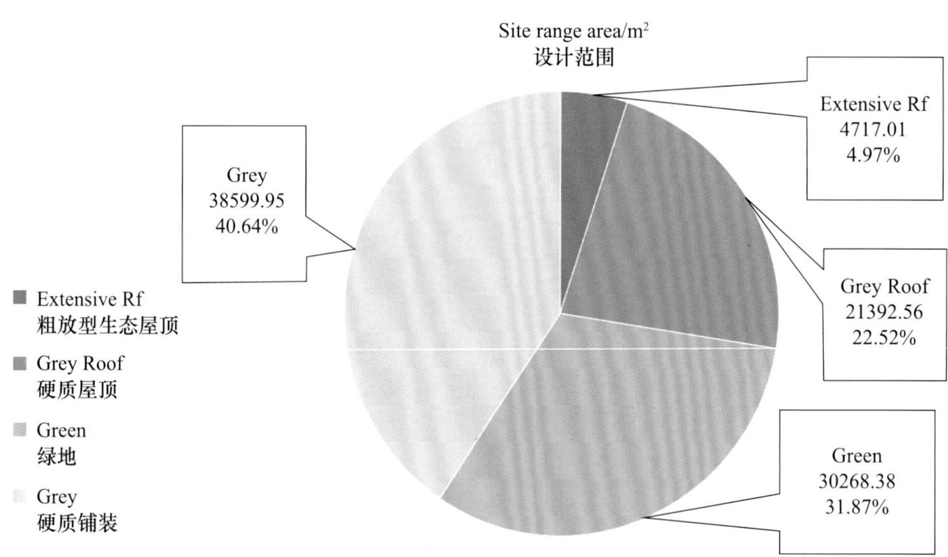

图 42 Pie chart of surface types and area within the site range
设计范围场地表面类型与面积饼状图

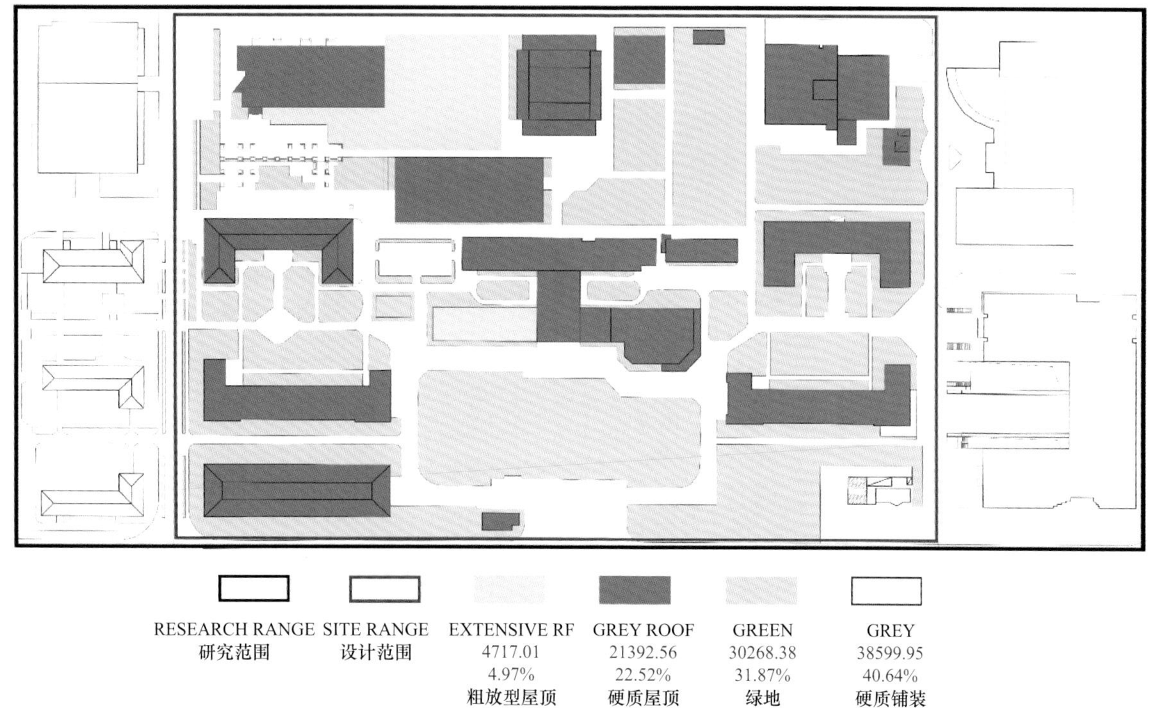

图 43　Schematic dragram of disfferent of surface types within the site range
设计范围内不同表面类型分布示意图

表 7　Calculation coefficients
计算系数

	面积	Roof Area 屋顶面积	Extensive Rf 粗放型屋顶	Intensive Rf 集约型屋顶	Grey Roof 硬质屋顶		面积	Ground Area 地面面积	Green 绿地	Grey 硬质铺装
Roof 屋顶	Area	26109.57	4717.01	0	21392.56	**Ground** 地面	Area	68868.33	30268.38	38599.95
	CN	—	—	—	98		CN	—	61	98
	Kc	—	0.3	0.6	—					

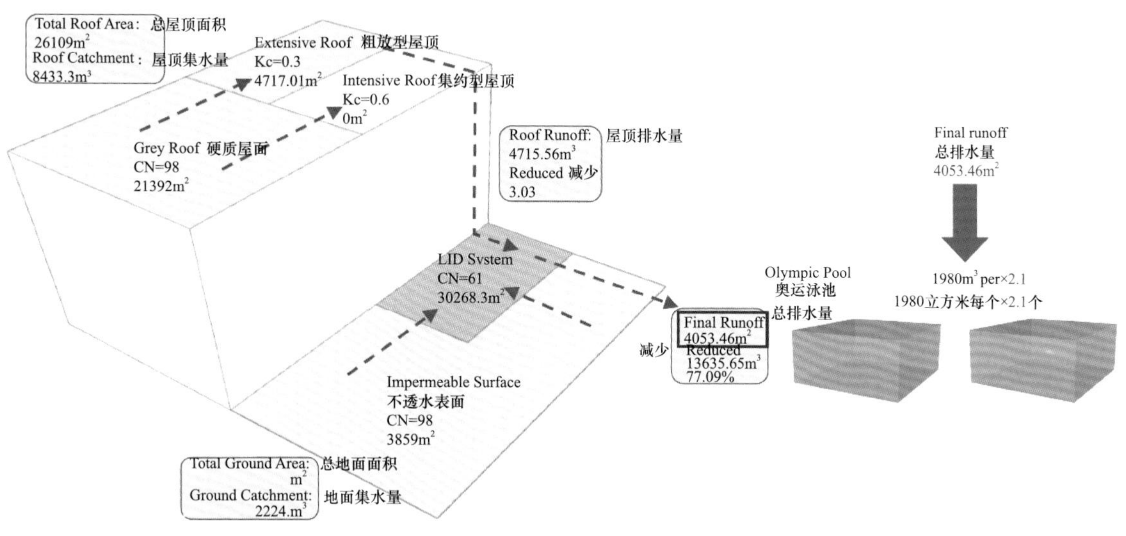

图 44　Calculation of runoff within the site range
设计范围场地径流量计算

2.3.2 Research Range 研究范围

表 8 Statistics of surface types and area within the research range
研究范围场地表面类型与面积统计

	The type of surface 表面类型	Research range/m^2 研究范围	Percentage/% 百分比
Roof 屋顶	Roof Area 屋顶面积	41537.02	—
	Extensive Rf 粗放型屋顶	4717.01	3.29%
	Intensive Rf 集约型屋顶	0	—
	Grey Roof 硬质屋顶	36820.01	25.72%
Ground 地面	Ground Area 地面面积	101627.91	—
	Green 绿地	35814.06	25.02%
	Grey 硬质铺装	65813.85	45.97%
	Sum 总和	143164.93	100.00%

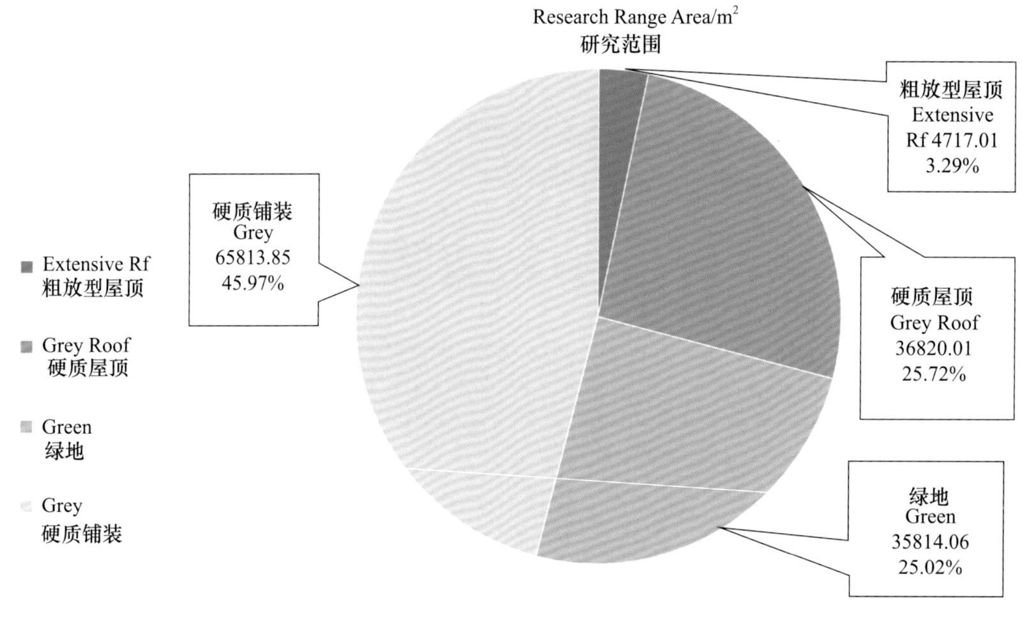

图 45 Pie chart of surface types and area within the research range
研究范围场地表面类型与面积饼状图

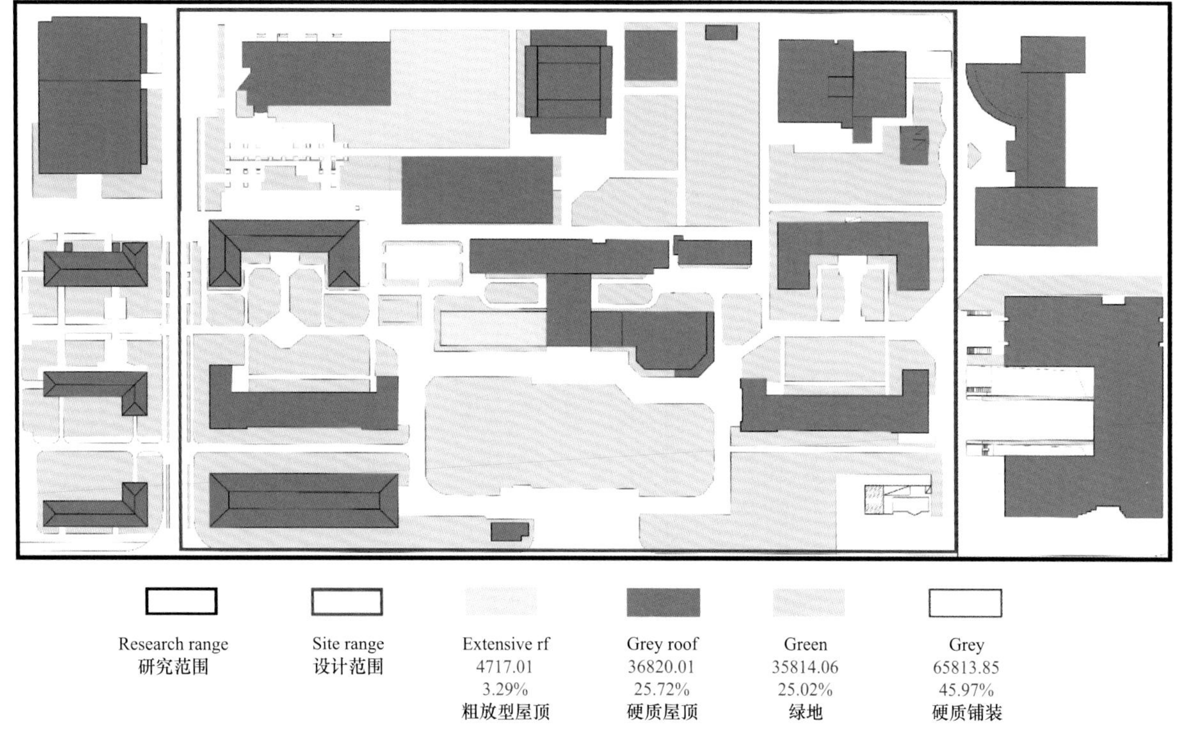

图 46 Schmatic diagram of distribution of diffareut surface types within tne research range
研究范围内不同表面类型分布示意图

表 9 Calculating coefficients
计算系数

	面积	Roof Area 屋顶面积	Extensive Rf 粗放型屋顶	Intensive Rf 集约型屋顶	Grey Roof 硬质屋顶		面积	Ground Area 地面面积	Green 绿地	Grey 硬质铺装
Roof 屋顶	Area	41537.02	4717.01	0	36820.01	**Ground** 地面	Area	101627.91	35814.06	65813.85
	CN	—	—	—	98		CN	—	61	98
	Kc	—	0.3	0.6	—					

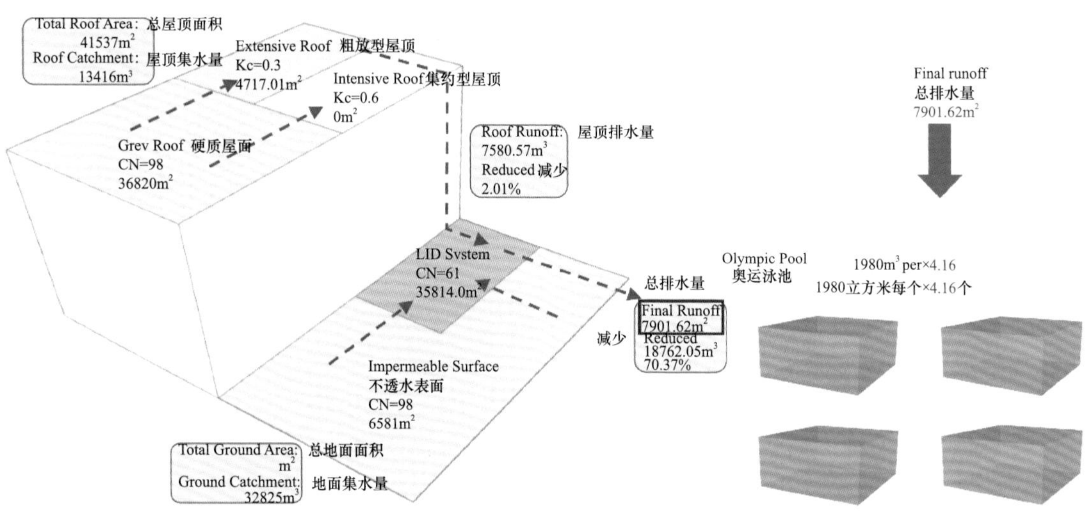

图 47 Calculation of runoff within the research range
研究范围场地径流量计算

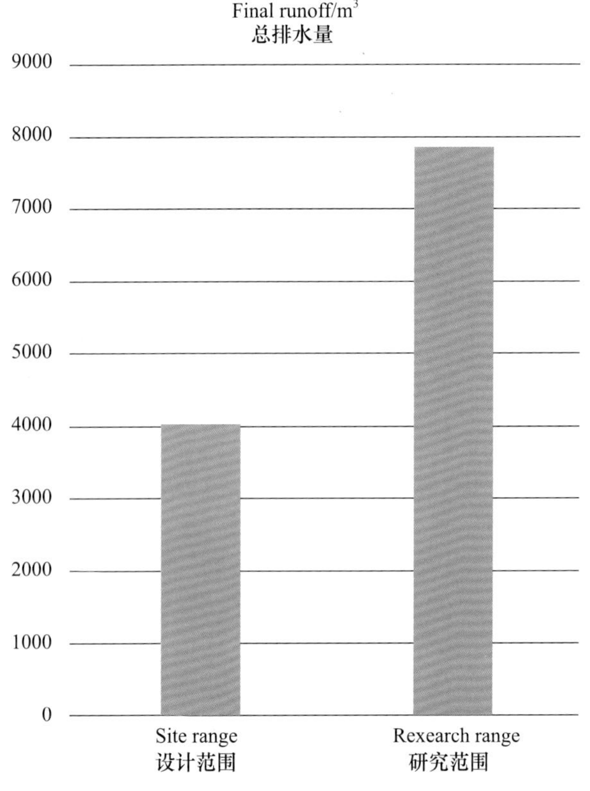

图 48　Comparison of final runoff between site range and research range
设计范围与研究范围总径流量比较

Considering the buildings and pavement around our site, there will be another 3848.16m^3 water which may flow to our site and need to be solved in our site. (as the worst situation)

考虑到我们场地周围的建筑物和路面，还有另外 3848.16m^3 的水可能会流入我们的场地，需要在我们的场地解决。(最糟糕的情况)

3. Design proposals 设计方案

3.1 Design of group one 第一组设计方案

3.1.1 Problem analysis 问题分析

图 49　Analysis of current problems
现状问题分析

1.The green slope in front of the building is high in the middle and low on both sides, which is not conducive to the storage and utilization of rainwater.

2. Curbstone blocks rainwater from seeping into green space.

1. 楼前绿地坡度中间高两侧低，不利于雨水储存利用。

2. 路缘石阻挡雨水渗入绿地。

3.1.2 Resolution strategies 解决策略

(1) Soften the trench 软化边沟设计

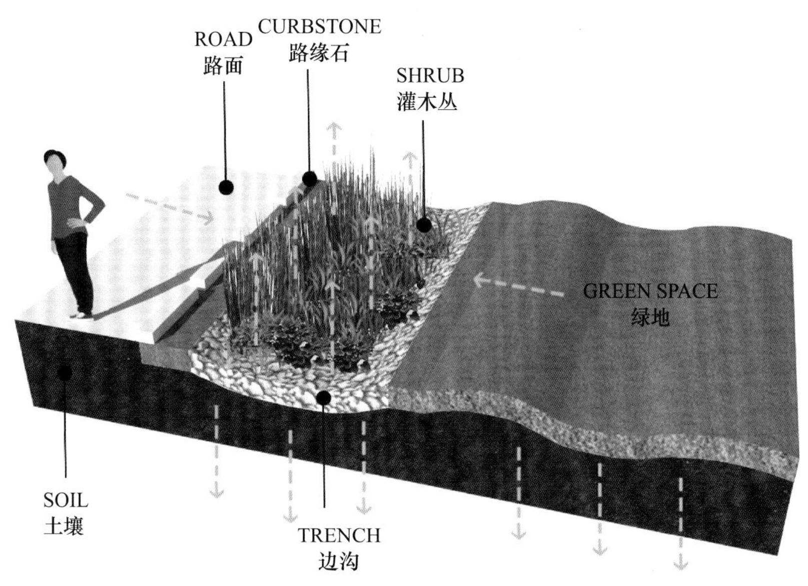

图 50 Planting shrubs in the trench
在边沟中种植灌木

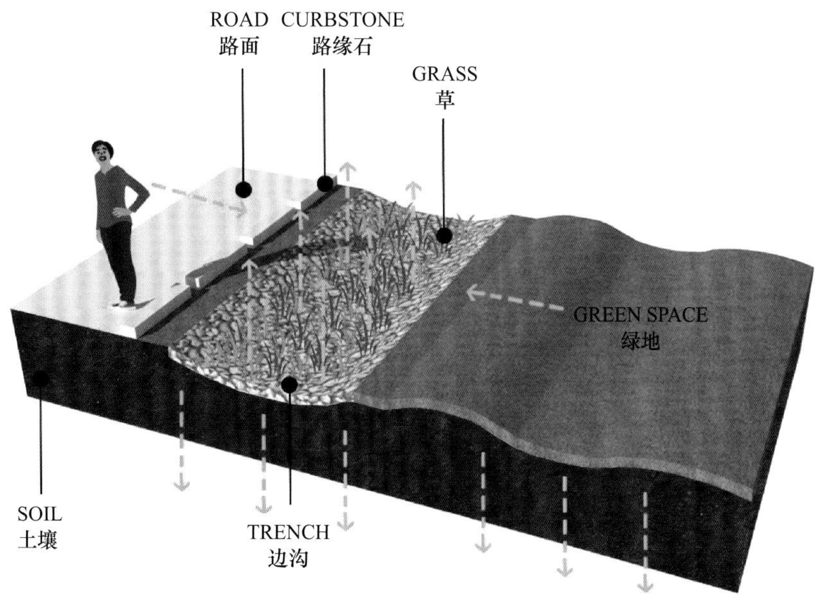

图 51 Planting grass in the trench
在边沟中植草

(2) Combination of Pavement Drainage and Trench Catchment 路面排水与边沟集水装置配合

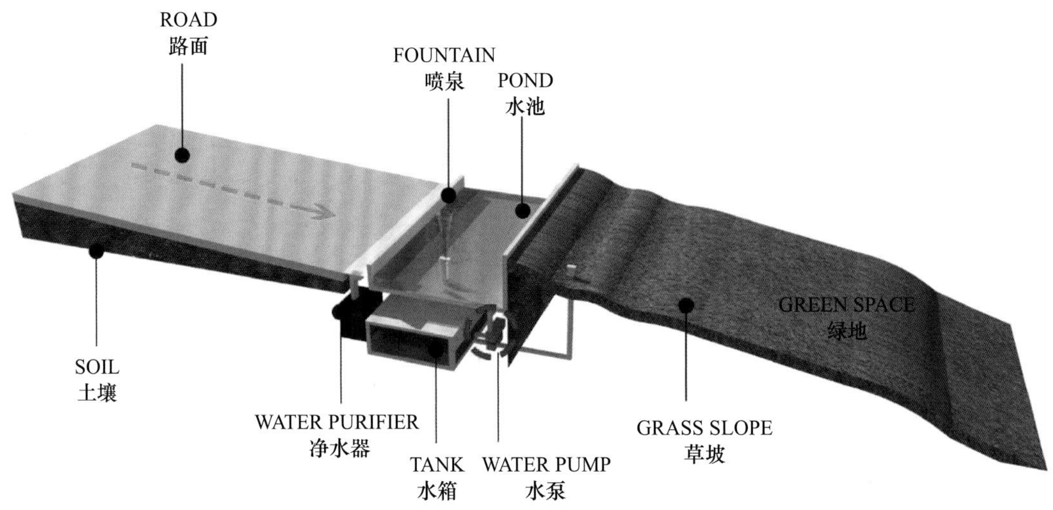

图 52　Collect rainwater from road to store water for landscaping
收集路面雨水储水造景

3.1.3 System construction 系统构建

(1) Green roof design 绿色屋顶设计

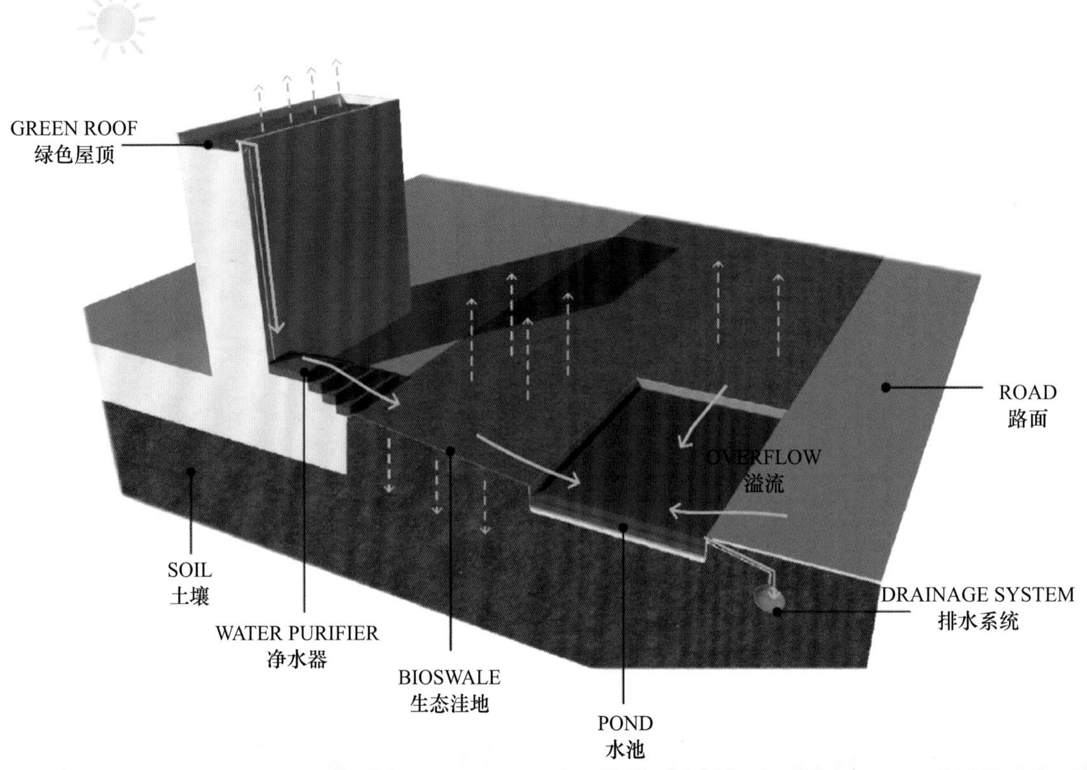

图 53　Green roof with waterscape
含水景的绿色屋顶

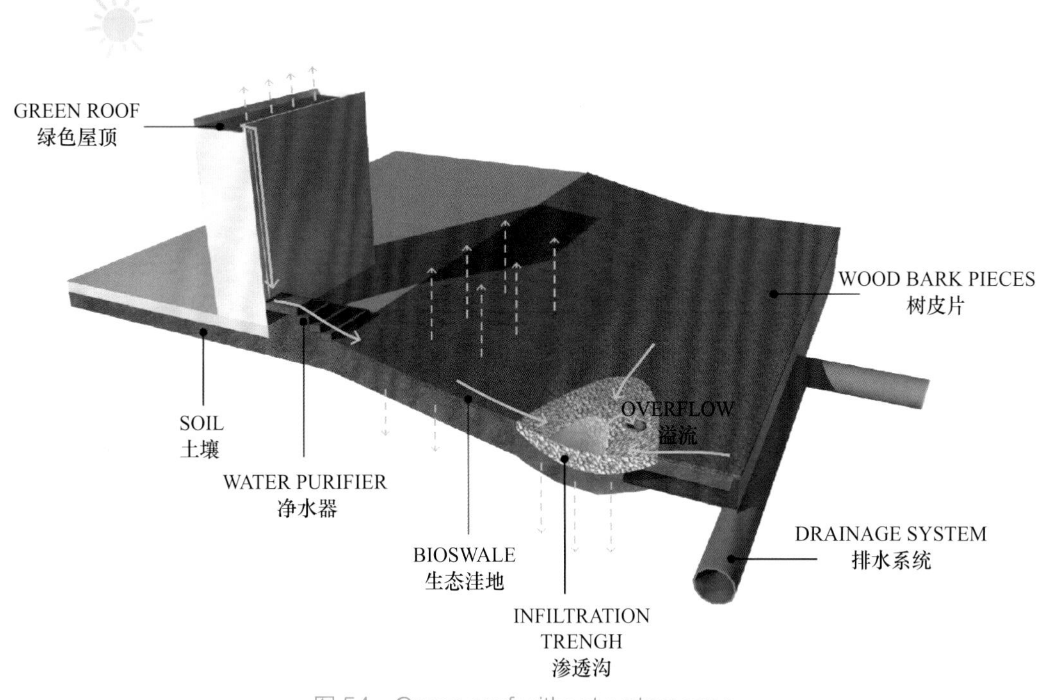

图 54　Green roof without waterscape
不含水景的绿色屋顶

(2) Gray roof design 硬质屋顶设计

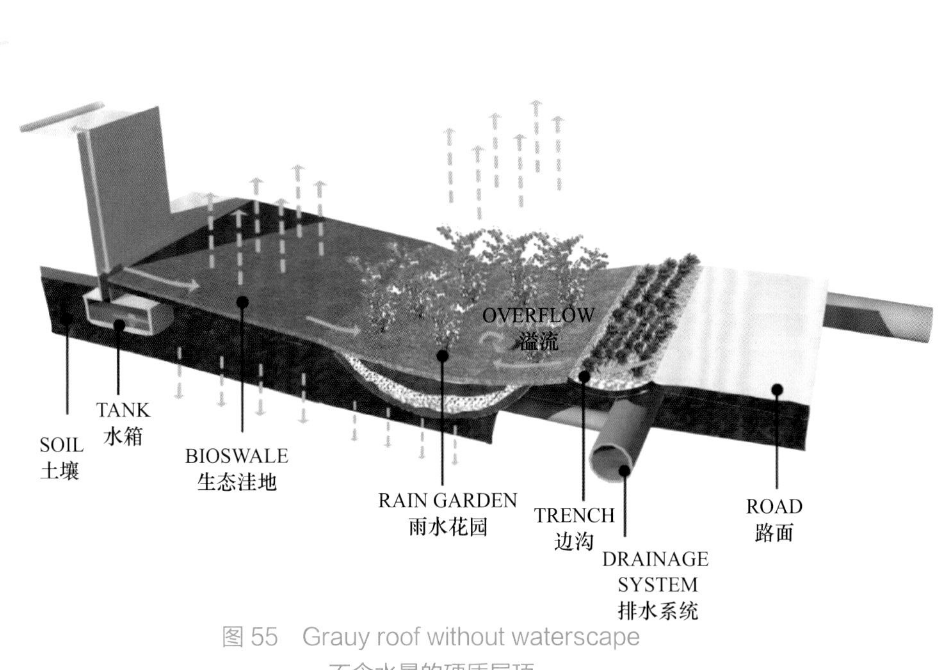

图 55　Grauy roof without waterscape
不含水景的硬质屋顶

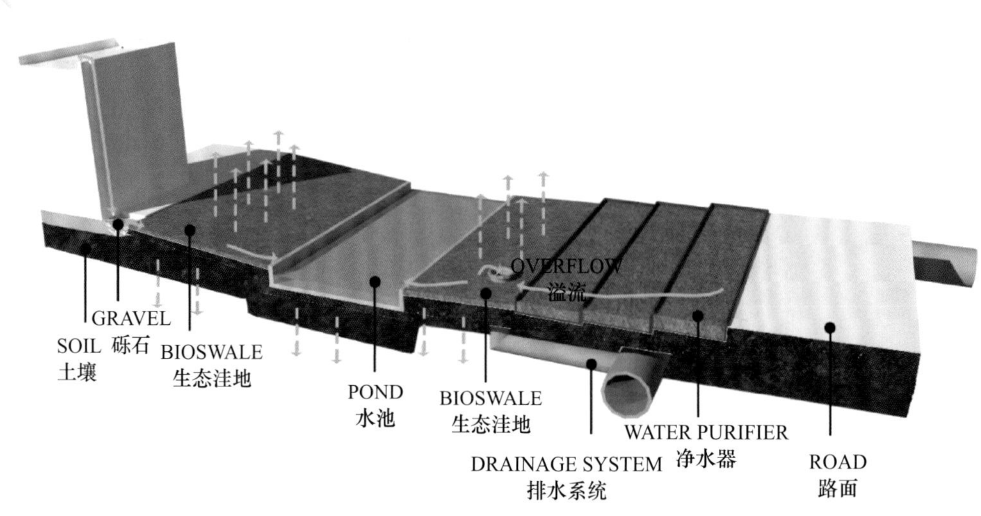

图 56 Grauy roof with waterscape
含有水景的硬质屋顶

3.1.4 Overall design 总体设计

图 57 Site design structure
场地设计结构

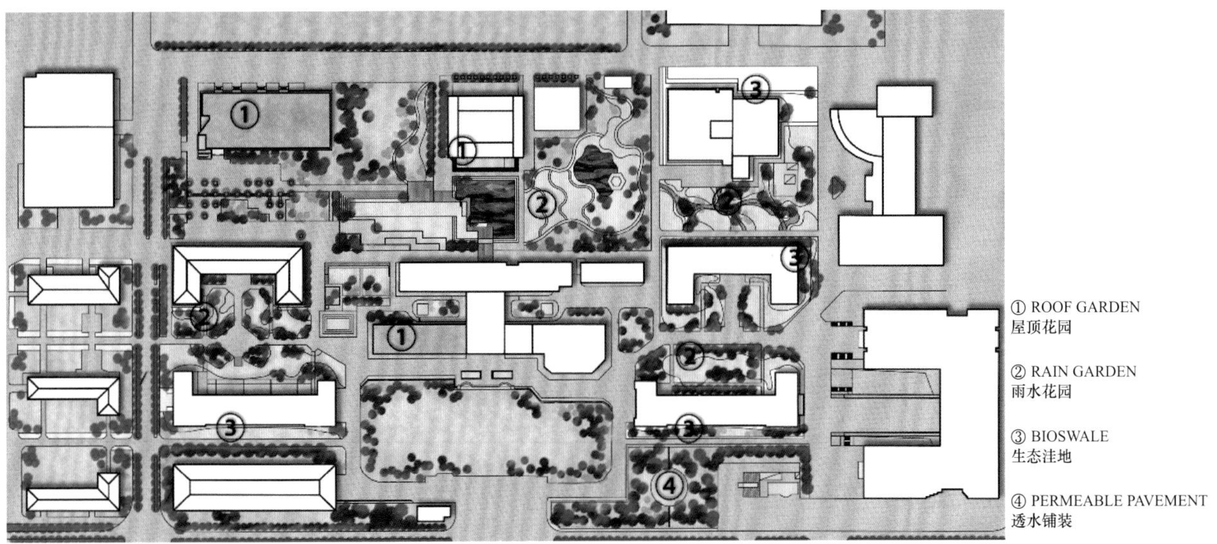

图 58 Master Plan
总平面图

3.1.5 Core area design 核心地段设计

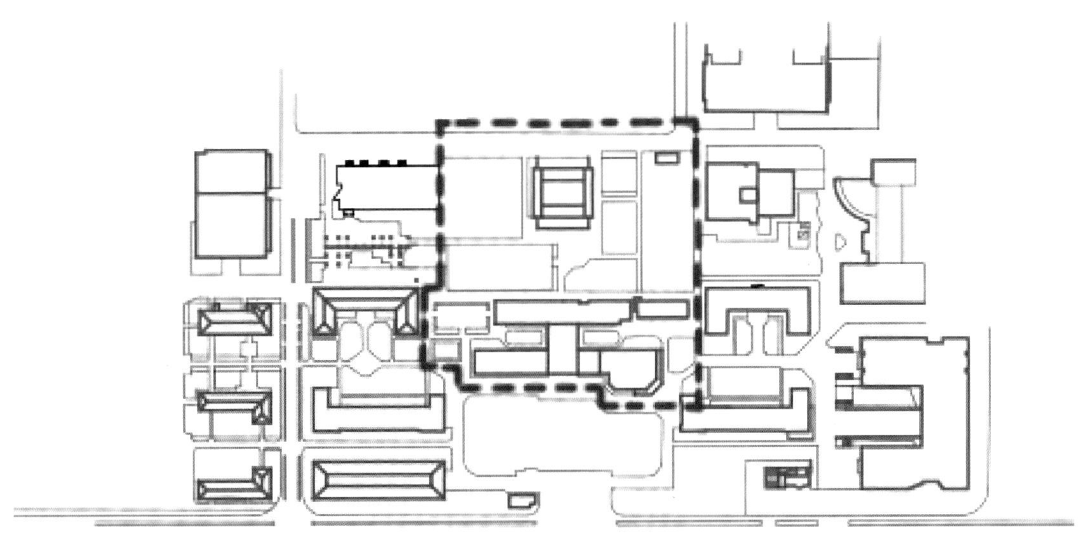

图 59 Location of the core area
核心地段位置示意

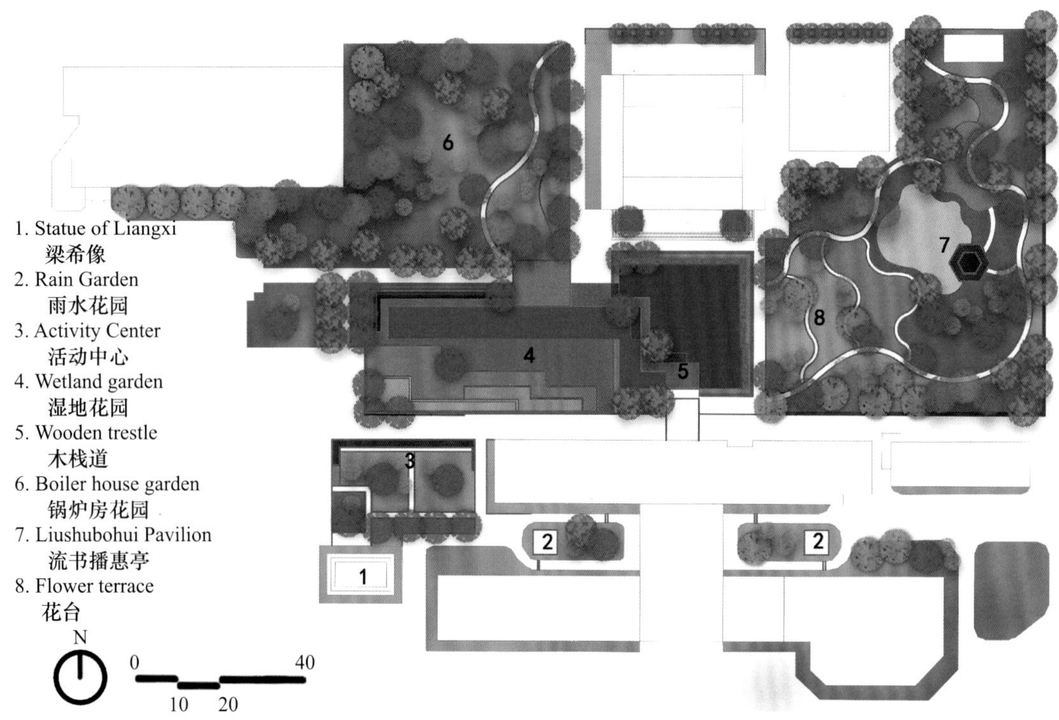

图 60 Plan of core area
核心地段平面图

图 61 Section1
剖面图一

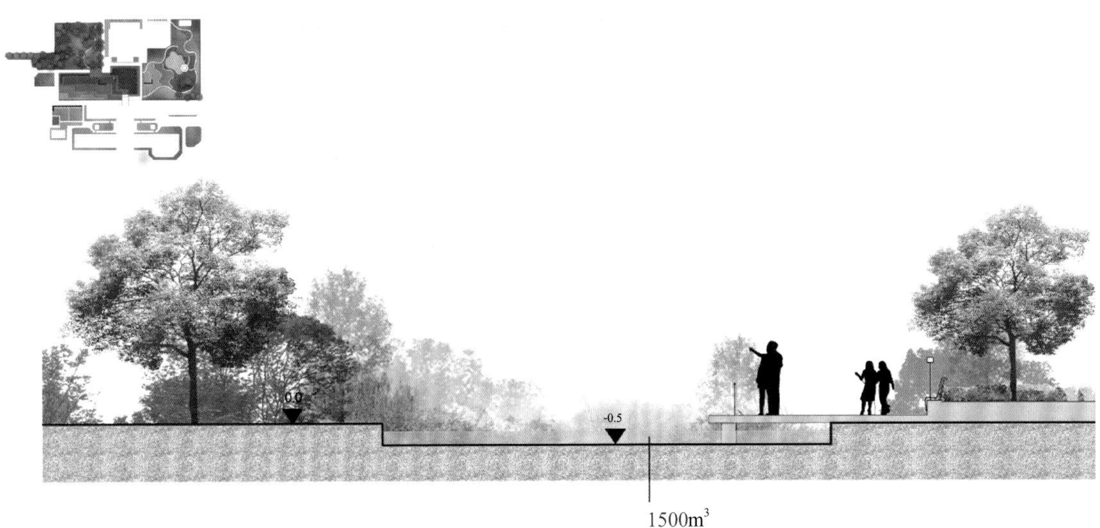

图 62 Section2
剖面图二

图 63 Section3
剖面图三

3.1.6 Other area design 其他区域设计

(1) Site 1 地块一

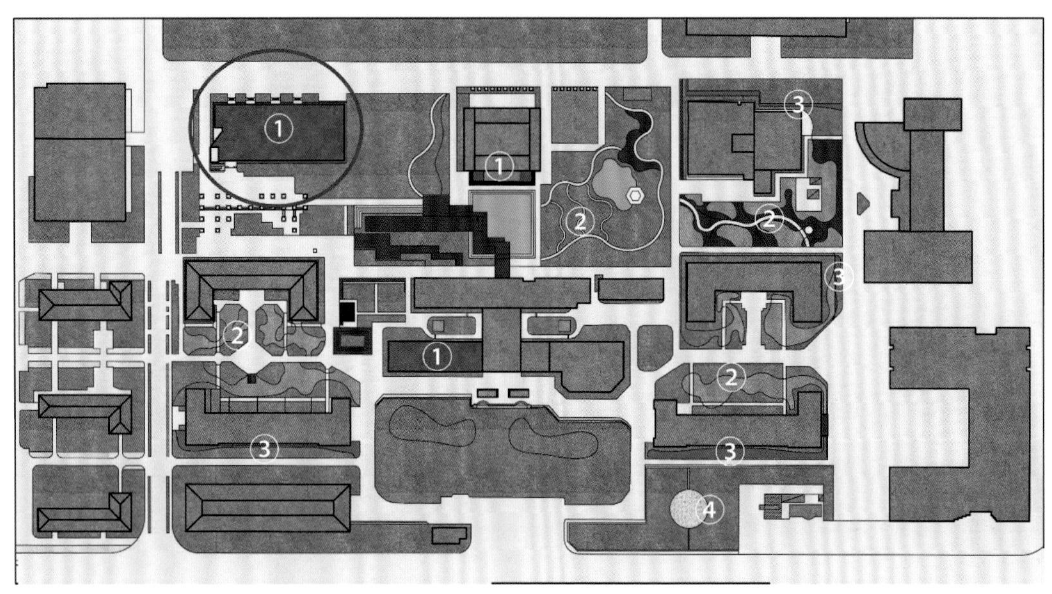

① ROOF GARDEN 屋顶花园　② RAIN GARDEN 雨水花园　③ BIO SWALE 生态洼地　④ PERMEABLE PAVEMENT 透水铺装

图 64　Location of site 1
地块一位置示意

图 65　Rainwater collection system
雨水收集系统

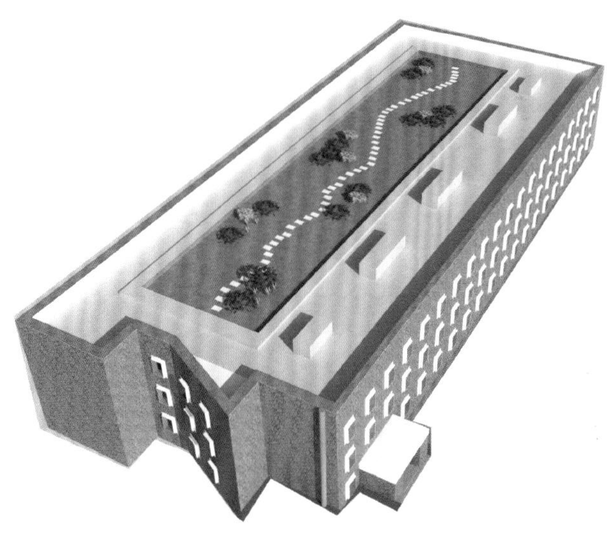

图 66　Green roof design
屋顶花园设计

(2) Site 2 地块二

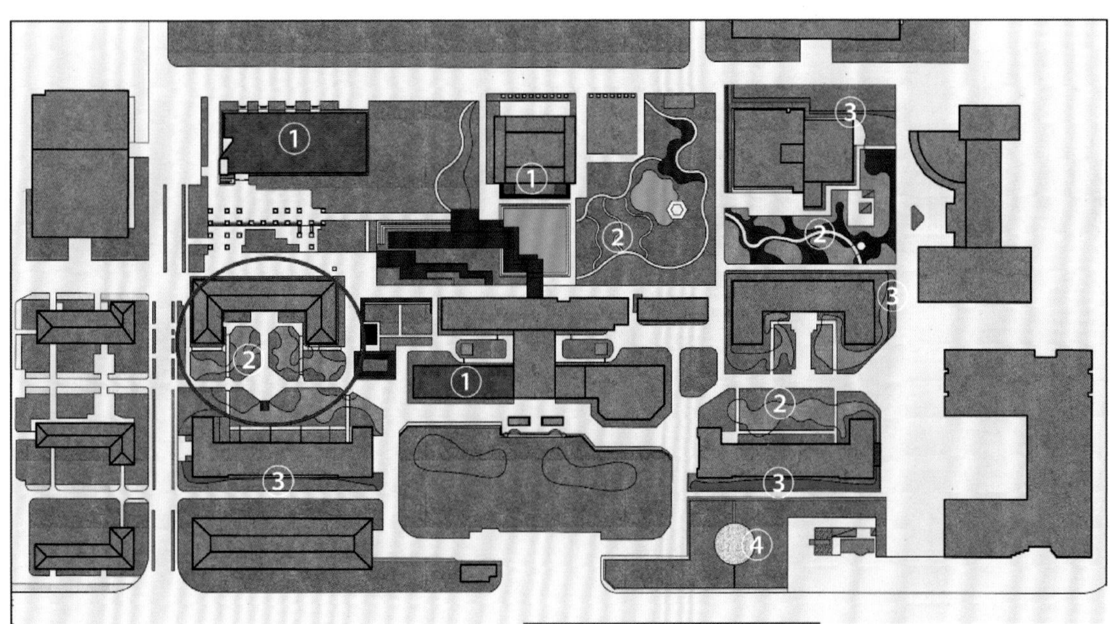

图 67　Location of site 2
地块二位置示意

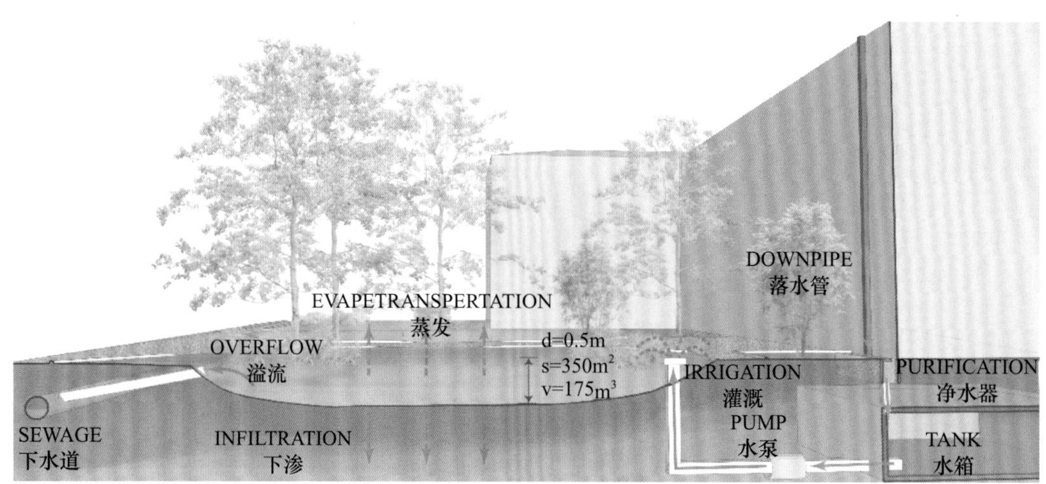

图 68 Rainwater collection system
雨水收集系统

(3) Site 3 地块三

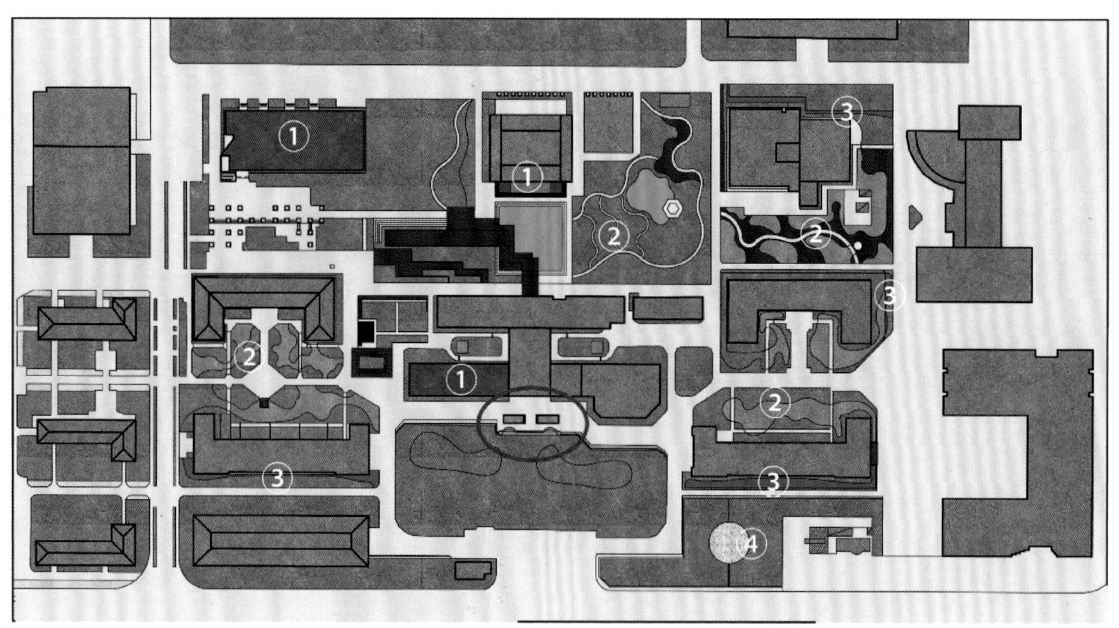

图 69 Location of site 3
地块三位置示意

图 70　Rainwater collection system
雨水收集系统

3.1.7 Suggestions on plant allocation 植物配置建议

Rain Garden Plant Schedule						
	Common Name	Latin Name	Type	Water consumption	Height	Spread
	Northern Sea Oats	*Chasmanthium latifolium*	Ornamental grass	💧💧💧	36"	24"
	Elijah Blue Fescue	*Festuca glauca 'Elijah Blue'*	Ground cover & Ornamental grass	💧	12"	10-15"
	German Iris	*Iris germanica*	Perennial	💧	4-40"	6-24"
	Silver Variegated Sweet Flag (Iris)	*Iris pallida 'Argentia-Variegata'*	Perennial	💧💧💧	24"	18"
	Feather Reed Grass	*Calamagrostis 'Karl Foerster'*	Ornamental grass	💧💧💧	60"	36"
	Japanese Pagoda Tree	*Sophora japonica*	Tree	💧💧	40-60'	30-45'
	Northern Gold Forsythia	*Forsythia 'Northern Gold'*	Shrub	💧💧	6-10'	6-12'
	Chokecherry	*Prunus virginiana*	Shrub	💧	12'	6'
	Vanderwolfe's Limber Pine	*Pinus flexis 'Vanderwolfe's'*	Tree	💧💧	26'	16'
	Globe Blue Spruce	*Picea pungens 'Globosa'*	Shrub	💧💧💧	4'	7'

图 71　Suggestions on plant allocation of rain gardens
雨水花园植物配置建议

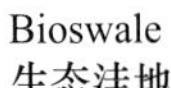
Bioswale
生态洼地

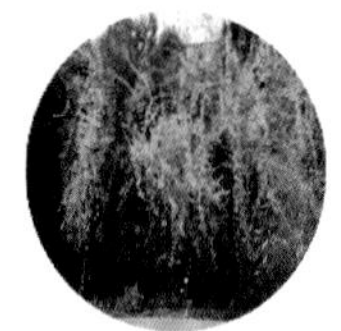
Northern Gold Forsythia
连翘

Chokecherry
美洲稠李

Wayfaring Tree
绵毛荚蒾

Terra Cotta Yarrow
蓍草

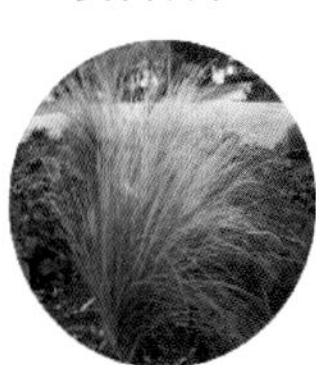
Mexican Feather Grass
墨西哥羽毛草

Junegrass
草地早熟禾

Bio Swale Plant Schedule

	Common Name	Latin Name	Type	Water consumption	Height	Spread
	Giant Chinese Silver Grass	*Miscanthus 'Giganteus'*	Ornamental grass	💧💧💧	40-60'	30-45'
	Elijah Blue Fescue	*Festuca glauca 'Elijah Blue'*	Ground cover & Ornamental grass	💧	12"	10-15"
	Idaho Fescue	*Festuca idahoensis*	Ground cover & Ornamental grass	💧💧	12"	12"
	Blue Hair Grass	*Viburnum lantana*	Ornamental grass	💧	12"	12"
	Feather Reed Grass	*Calamagrostis 'Karl Foerster'*	Ornamental grass	💧💧💧	60"	36"
	Golden Tufted Hair Grass	*Deschampsia flexuosa 'Aurea'*	Ornamental grass	💧	8-18"	12"
	Heavy Metal Switch Grass	*Panicum virgatum 'Heavy Metal'*	Ornamental grass	💧💧	54"	36"
	Junegrass	*Koeleria cristata (macrantha)*	Ornamental grass	💧	18"	12"
	Mexican Feather Grass	*Stipa tenuissima*	Ornamental grass	💧	12-18"	12-18"
	Variegated Japanese Sedge	*Carex oshimensis 'Evergold'*	Ornamental grass	💧💧💧	12"	18"

图 72　Suggestions on plant allocation of bioswales
生态洼地植物配置建议

3.1.8 Design Effect Accounting 设计效果核算

(1) Chinese calculation method 中式算法

$$W=10\psi_{ZC}h_yF$$

W——The total runoff (m^2) 总径流量；

ψ_{ZC}——Rainfall comprehensive runoff coefficient 雨水综合径流系数；

h_y——Designed rainfall (mm) 设计雨量；

F——Catchment area (hm^2) 集水面积。

① Surface alteration 表面材质变动

Adding 增加	Area (m^2) 面积
green land 绿地	2330.36
green roof 绿色屋顶	3079.36
pavement (pervious) 透水铺装	811.68
water surface 水面	1165.98

Cutting 减少	Area (m^2) 面积
pavement (impervious) 不透水铺装	4308.02
grey roof 硬质屋顶	5515.08

We also have designed water tanks to store about 1800m^3 water.

我们还设计了水箱，可存储约 1800 立方米的水。

BEFORE
设计前

GREY 63.16%
硬质

GREEN 36.84%
绿化

AFTER
设计后

GREY 52.83%
硬质

GREY 47.17%
硬质

② Calculation of site range 设计范围核算

表 10 Statistics of surface types and area within the site range (after design)
设计范围场地表面类型与面积统计（设计后）

The type of surface 表面类型	**Area** (m^2) 面积	**Percentage** (%) 百分比
green roof 生态屋顶	3893.97	4.10%
pavement (impervious) 不透水铺装	34291.93	36.11%
pavement (pervious) 透水铺装	811.68	0.85%
green land 绿地	32598.74	34.32%
water surface 水面	1165.98	1.23%
grey roof 硬质屋顶	15877.48	16.72%
basement which is covered with green land 绿地覆盖的地下室	6338.12	6.67%
sum 总和	94977.90	1.00

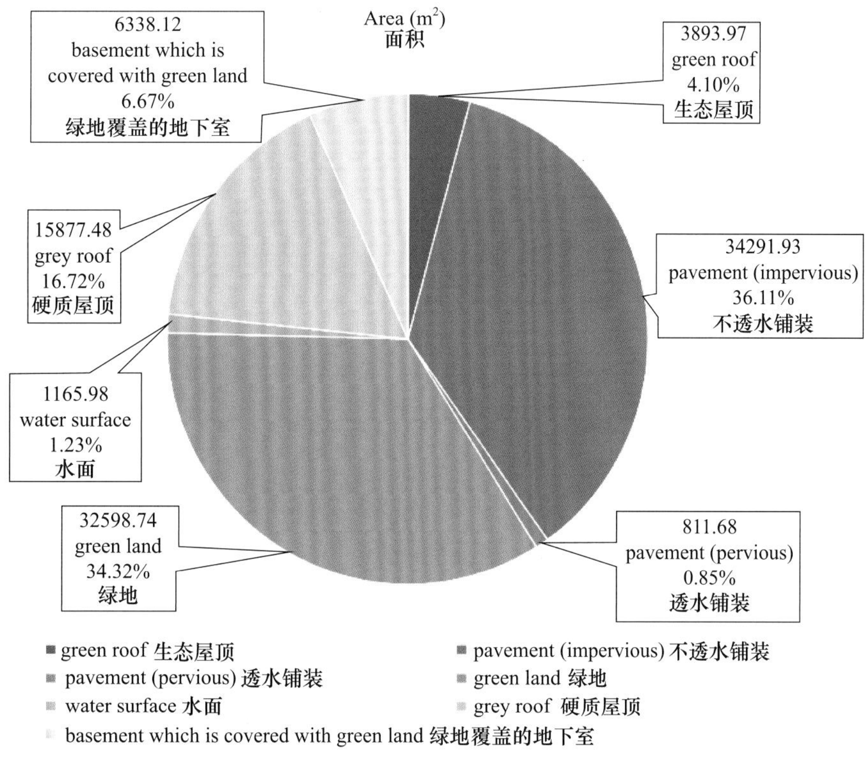

图 73 Pie chart of surface types and area within the site range (after design)
设计范围场地表面类型与面积饼状图（设计后）

表 11　Calculation of final runoff in site range (after design)
设计范围总径流量计算（设计后）

Designed (site range) 设计后（设计范围）	Area (m^2) 面积	Runoff coefficient 径流系数	W (runoff/m^3) 总径流量	
Type 类别			10 year period（10 年期）	3 year period（3 年期）
green roof 生态屋顶	3893.97	0.30	244.15	126.16
pavement (impervious) 不透水铺装	34291.93	0.90	6540.31	3333.18
pavement (pervious) 透水铺装	811.68	0.25	42.41	21.92
green land 绿地	32598.74	0.15	1021.97	528.10
water surface 水面	1165.98	1.00	243.69	125.93
grey roof 硬质屋顶	15877.48	0.90	2986.55	1543.29
basement which is covered with green land 绿化覆盖的地下室	6338.12	0.15	198.70	102.68
sum 总和	94977.90	—	11280.78	5781.26

Minus water stored in the tanks (1800m^3)
Final runoff is 3981.25m^3
减去存储于水箱中的1800m^3水
最终径流量为3981.25m^3

=Olympic Pool 1980m^3 per×2.1
奥运泳池1980立方米每个×2.1个

图 74　Schematic diagram of fiual runoff within the site range (after design)
设计范围场地总径流量示意（设计后）

③ Calculation of research pange 研究范围核算

表 12　statistics of surface types and area within the research range　(after design)
研究范围场地表面类型与面积统计（设计后）

The type of surface 表面类型	Area (m^2) 面积	Percentage (%) 百分比
green roof 生态屋顶	3893.97	2.72%
pavement (impervious) 不透水铺装	61502.83	42.96%
pavement (pervious) 透水铺装	811.68	0.57%
green land 绿地	38147.42	26.64%
water surface 水面	1165.98	0.81%
grey roof 硬质屋顶	31304.93	21.87%
basement which is covered with green land 绿地覆盖	6338.12	4.43%
sum 总和	143164.93	1.00

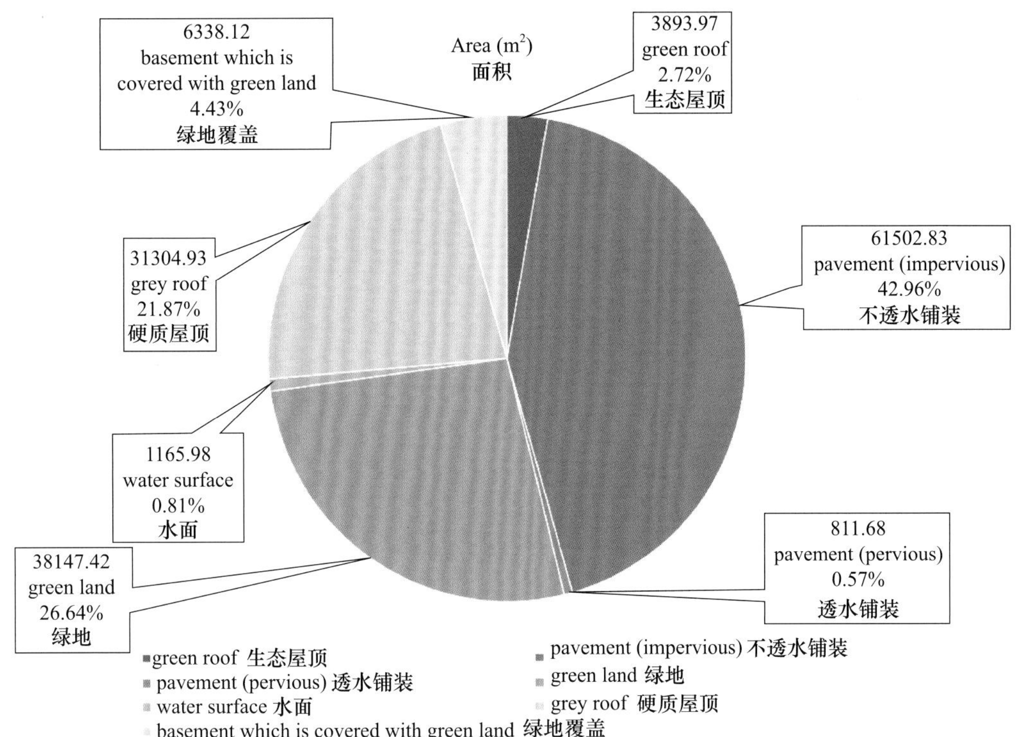

图 75 Pie chart of surface types and area within the research range (after design)
研究范围场地表面类型与面积饼状图（设计后）

表 13 Calculation of fiual runoff in research range (after design)
研究范围总径流量计算（设计后）

Designed (research range) 设计后（研究范围）	**Area (m²)** 面积	**Runoff coefficient** 径流系数	**W (runoff/m³)** 总径流量	
Type 类别			**10 year period**（10 年期）	**3 year period**（3 年期）
green roof 生态屋顶	3893.97	0.30	244.15	126.16
pavement (impervious) 不透水铺装	61502.83	0.90	11568.68	5978.08
pavement (pervious) 透水铺装	811.68	0.25	42.41	21.92
green land 绿地	38147.42	0.15	1195.92	617.99
water surface 水面	1165.98	1.00	243.69	125.93
grey roof 硬质屋顶	31304.93	0.90	5888.46	3042.84
basement which is covered with green land 绿地覆盖的地下室	6338.12	0.15	198.70	102.68
sum 总和	143164.93	—	19382.01	10015.60

Minus water stored in the tanks (1800m^3)
Final runoff is 8215.59m^3
减去存储在水箱中的1800m^3水
最终径流量为8215.59m^3

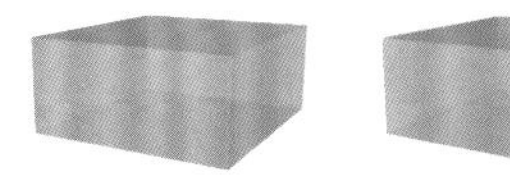

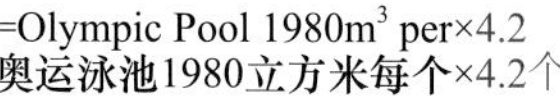

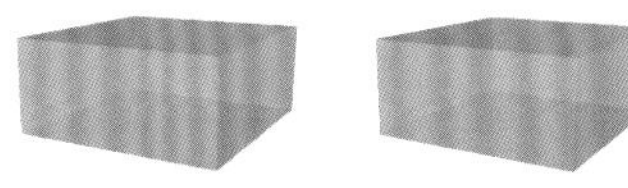

图 76 Schematic diagram of final runoff within the research range (after design)
研究范围场地总径流量示意（设计后）

表 14 Final runoff analysis of site range and research range before and after design
设计前后设计范围与研究范围总径流量分析

	W (runoff/m^3) 总径流量		Comparison 差值	Percentage (%) 百分比
	Before 设计前	After 设计后		
siter range 设计范围	6411.23	3981.25	2429.98	37.90%
research range 研究范围	10645.57	8215.59	2429.98	22.83%

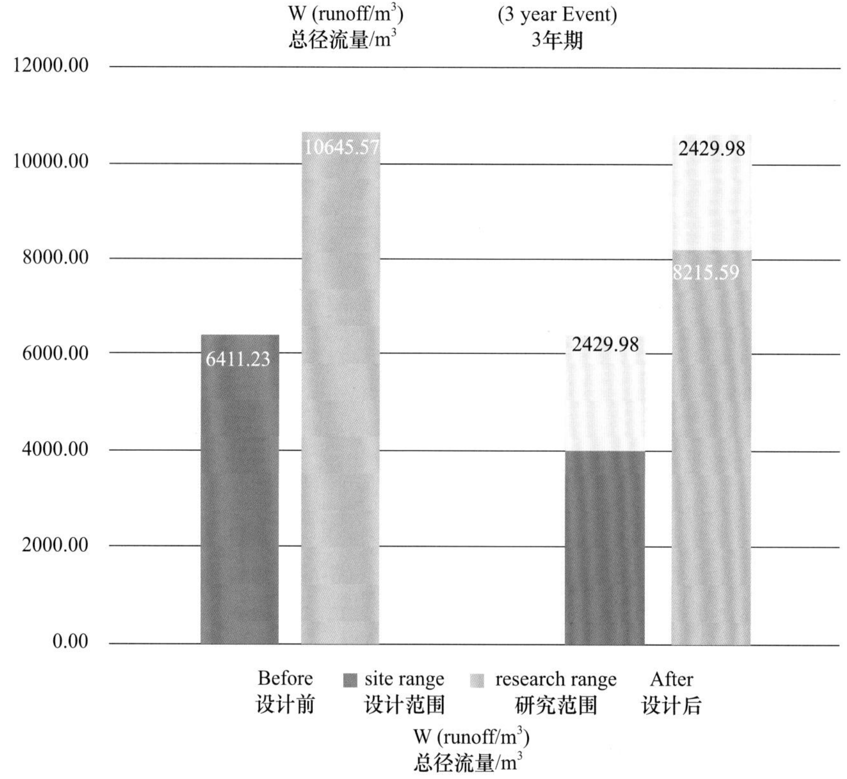

图 77 Comparison of final runoff between site range and research range before and after design
设计前后设计范围与研究范围总径流量对比

Since we have to consider the situation of 3 year period when we do our design, we take the calculation of 3 year period and research range for analysis and the calculation of 10 year period and site range for reference.

由于在设计的时候必须考虑 3 年期间的状况，所以我们采用 3 年期的计算和研究范围进行分析，计算 10 年期和现场范围，以供参考。

After the design, we successfully reduce the runoff of our site by 2429.98m^3 (22.83%) (as for research range).

设计完成后，我们成功减少了 2429.98m^3（22.83%）的场地径流量（研究范围）。

Thus, the water we save can be used by an adult for about 64 years.

因此，我们节约的水可以由一个成年人使用约 64 年。

(2) UBC Calculation method UBC 算法

① surface alternation 表面材质变动

表 15 Surface alternation of the site
场地表面材质变动

Adding 增加	Area (m^2) 面积
Extensive Rf 粗放型屋顶	5515.08
Green 绿地	2330.36

Cutting 减少	Area (m^2) 面积
Grey Roof 硬质屋顶	5515.08
Grey 不透水铺装	2330.36

We also have designed water tanks to store about 1800m^3 water.

我们还设计了水箱，可存储约 1800 立方米的水。

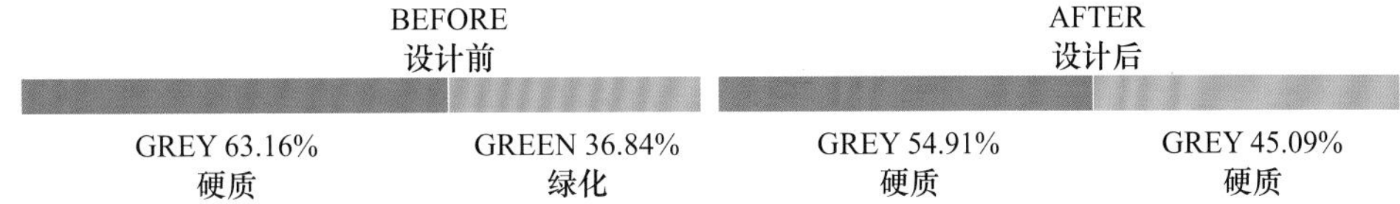

② Calculation of site range 设计范围核算

表 16 Statistics of surface types and area within the site range（after design）
设计范围表面类型与面积统计（设计后）

	The type of surface 表面类型	Site range/m^2 设计范围	Percentage/% 百分比
Roof 屋顶	Roof Area 屋顶面积	26109.57	—
	Extensive Rf 粗放型屋顶	10232.09	10.77%
	Intensive Rf 集约型屋顶	0.00	—
	Grey Roof 硬质屋顶	15877.48	16.72%
Ground 地面	Ground Area 地面面积	68868.33	—
	Green 绿地	32598.74	34.32%
	Grey 铺装	36269.59	38.19%
	Sum 总和	94977.90	100.00%

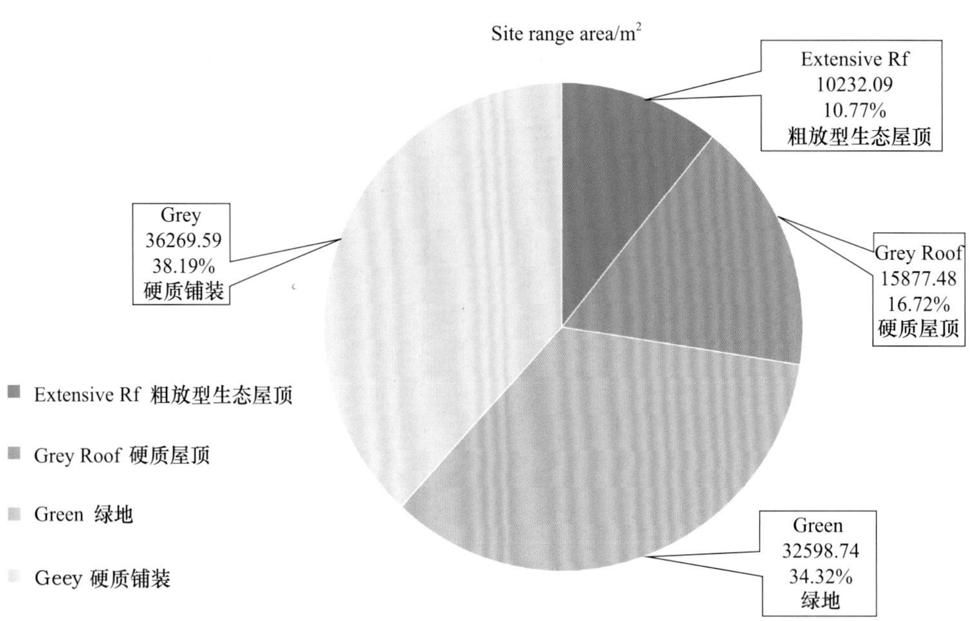

图 78 Pie chart of surface types and area within the site range (after design)
设计范围场地表面类型与面积饼状图（设计后）

表 17 Calculation coefficients
计算系数

	面积	**Roof Area** 屋顶面积	**Extensive Rf** 粗放型屋顶	**Intensive Rf** 集约型屋顶	**Grey Roof** 硬质屋顶		面积	**Ground Area** 地面面积	**Green** 绿地	**Grey** 硬质铺装
Roof 屋顶	Area	26109.57	10232.09	0	15877.48	**Ground** 地面	Area	68868.33	32598.74	36269.59
	CN	—	—	—	98		CN	—	61	97.26
	Kc	—	0.3	0.6	—					

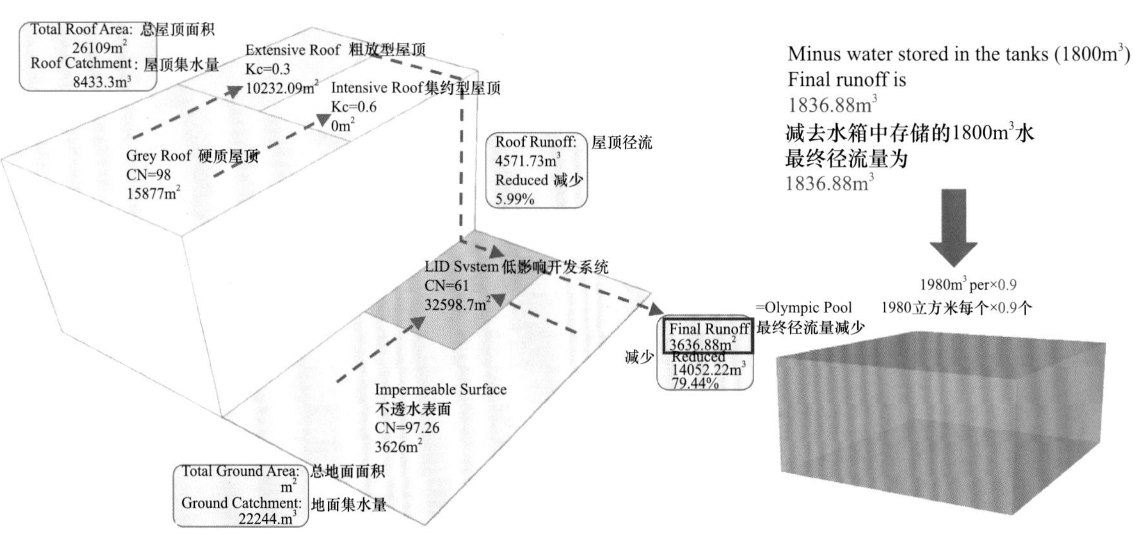

图 79 Calculation of runoff within the site range
设计范围场地径流量计算（设计后）

③ Calculation of research range 研究范围核算

表 18 Statistics of surface types and area within the research range (after design)
研究范围表面类型与面积统计（设计后）

	The type of surface 表面类型	Site range/m² 设计范围	Percentage/% 百分比
Roof 屋顶	Roof Area 屋顶面积	41537.02	—
	Extensive Rf 粗放型屋顶	10232.09	7.15%
	Intensive Rf 集约型屋顶	0	—
	Grey Roof 硬质屋顶	31304.93	21.86%
Ground 地面	Ground Area 地面面积	101627.91	—
	Green 绿地	38147.42	26.65%
	Grey 铺装	63480.49	44.34%
	Sum 总和	143164.93	100.00%

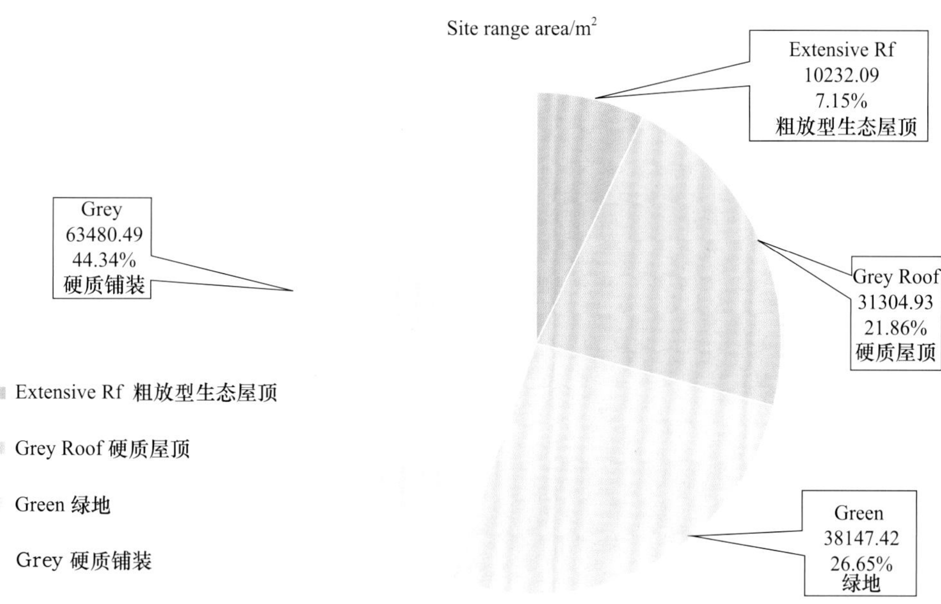

图 80 Pie chart surface types and area within the research range (after design)
研究范围场地表面类型与面积饼状图（设计后）

表 19 Calculation coefficients
计算系数

	面积	Roof Area 屋顶面积	Extensive Rf 粗放型屋顶	Intensive Rf 集约型屋顶	Grey Roof 硬质屋顶		面积	Ground Area 地面面积	Green 绿地	Grey 硬质铺装
Roof 屋顶	Area	41537.02	10232.09	0	31304.93	Ground 地面	Area	101627.91	38147.42	63480.49
	CN	—	—	—	98		CN	—	61	97.58
	Kc	—	0.3	0.6	—					

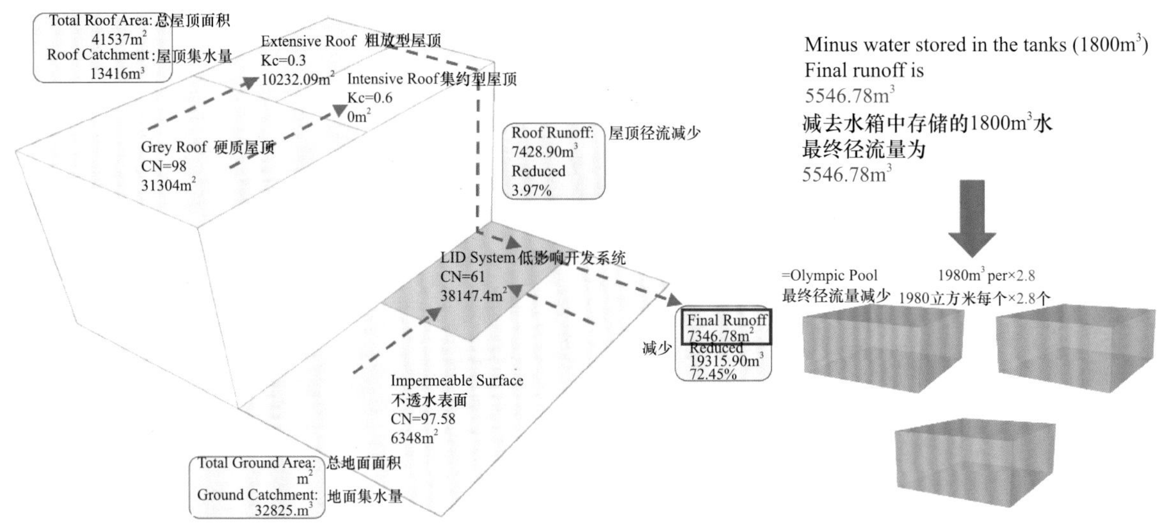

图 81 Calculation of runoff within the research range
研究范围场地径流量计算

表 20 Final runoff analysis of site range and research range before and after design
设计前后设计范围与研究范围最终径流量分析

	Final runoff/m³ 最终径流量		Comparison 差值	Percentage (%) 百分比
	Before 设计前	**After** 设计后		
siter range 设计范围	4053.46	1836.88	2216.58	54.68%
Research range 研究范围	7901.62	5546.78	2354.84	29.80%

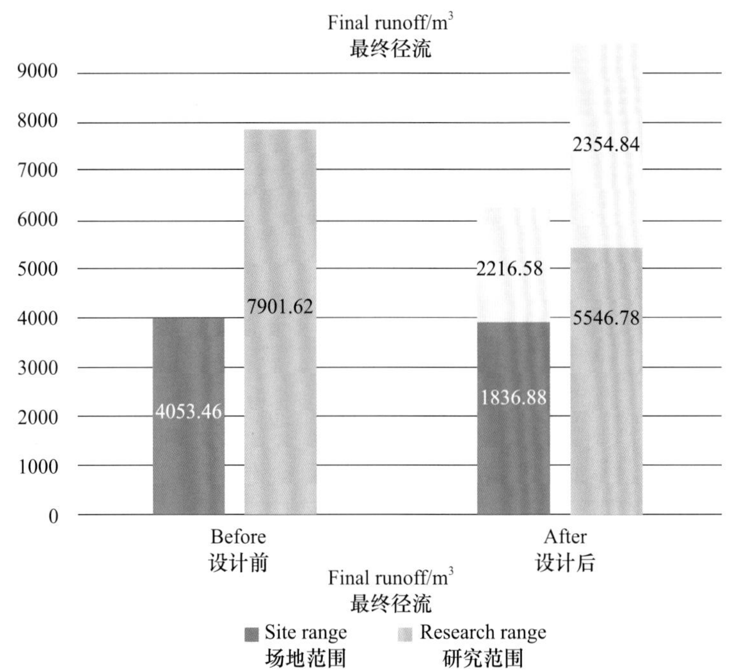

图 82 Comparison of final runoff between site range and research range before and after design
设计前后设计范围与研究范围最终径流量对比

After the design, we successfully reduce the runoff of our site by 2354.84m^3 (29.80%) (as for research range).

Thus, the water we save can be used by an adult for about 62 years.

设计完成后，我们成功减少了场地的径流量 2354.84m^3（29.80%）（研究范围）。

因此，我们节约的水可以由一个成年人使用约 62 年。

3.1.9 Design cost calculation 设计成本核算

表 21 Transformation costs
改造成本

Facility type 设施类型	Amount 数量	Unit price 单价	Total/RMB 总价
green land 绿地	3476.34	30.00	104290.20
green roof 生态屋顶	3079.36	150.00	461904.00
pavement (pervious) 透水铺装	811.68	100.00	81168.00
water surface 水面	1165.98	350.00	408093.00
water tank 水箱	1800.00	800.00	1440000.00
bioswale 生态洼地	1086.72	40.00	43468.80
sunken green space 下沉绿地	5618.92	40.00	224756.80
sum 总和	17039.00	—	2763680.8

3.1.10 Design proposals 设计提议

THE BENEFITS OF LID
低影响开发的优点

Rainfall
降水

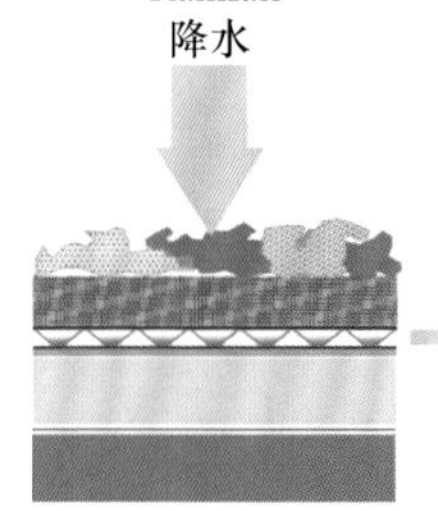

runoff
径流

-Runoff reduction
-Delay in peak flow
降少径流量
推迟洪峰

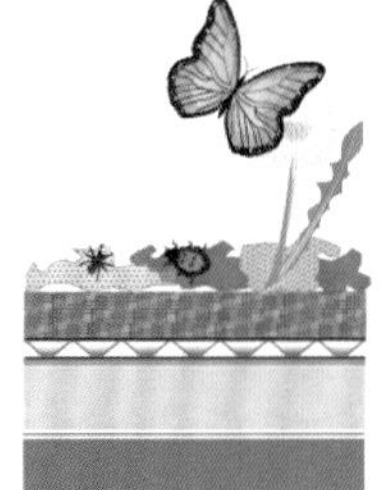

-Bio-diversity
-Amenity and aesthetic
(提高) 生物多样性
(提高) 舒适性与美观性

Money saving
节约资金

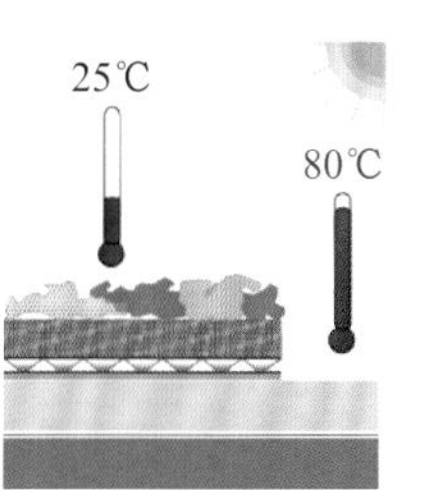

-Building envelope protection
-Extend the life span
保护建筑表面
延长寿命

filter out
dust particles
过滤粉尘

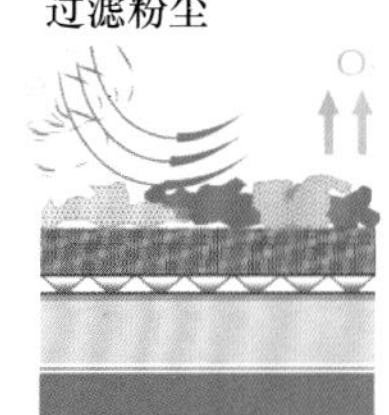

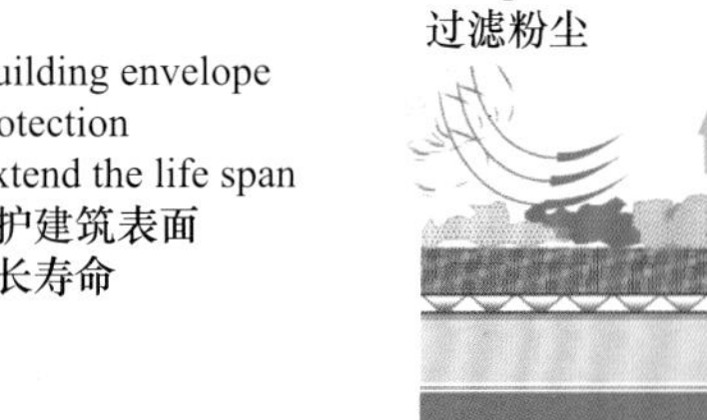

-Air quality
-Microclimate
(提高) 空气质量
(改善) 微气候

Education
教育功能

HOW TO PERSUADE THE PUBLIC
如何说服民众

SHORT VIDEOS
短片

MEETINGS WITH COMMUNITY MEMBERS
与社区代表见面

PUBLICATIONS
出版物

2017 Roehr, D., 2017,“ Keep the Arbutus Corridor for Cyclist and Walkers. ”the Tyee.ca.(Link)
Luo, C., Roehr, D., 2017, “People's Use of Urban Small Parks: A case Study of Haidian, Beijing, China. ” Bridging . CELA. China, Beijing, (Digital proceedings)

2016 Roehr D., Fassman-Beck, E., 2016. “Regenwasserrueckhaltung-warum Dachbegruenung weiterhin ein Thema der Zukunft ist”. Neue Landschaft(11): 41-48
Roehr D., Sjoquist, M., 2016. “Social Structures”. Topos (94): 70-77.

LECTURES FOR STUDENTS
学生讲座

REPORTS FOR CHAIRMAN
向负责人汇报

3.1.11 Future imagination 未来畅想

图 83 Landscape between teaching buildings
教学楼间景观

图 84　Landscape of students' square
学生广场景观

图 85　Landscape of “Liuyun” Pavilion
流云亭景观

3.2 Design proposal of group two 第二组设计方案

3.2.1 Site 1 地块一设计

site present 场地现状

图 86 Current situation of the site
场地现状

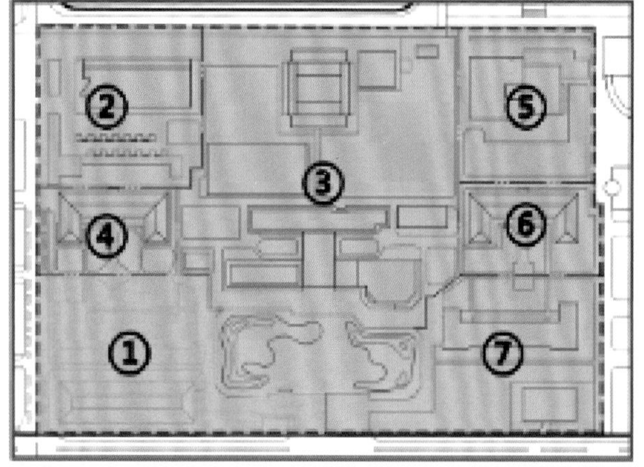

图 87 Location of Sitel
地块一位置示意

Function 主要功能：
Attached green space
建筑附属绿地
Appreciate lawn
观赏草坪
Isolation of external noise
隔绝外部噪声
Green space matching with building facade
配合建筑立面的景观绿地

Advantage 优点：

1. Already have LID auareness 已有 LID 的意识

2. There are stones and drainage ditchs on the roadside, and small rainwater gardens ave set in the green space 路边有石子以及排水沟，绿地中设置

disadvantage 缺点：

1. The ground bulges and cannot collect water. 美桐大道行道树过大，地面凸起无法集水

2. Rain gardens are below the drain and plants do not grow well. 雨水花园低于排水口，且植物生

小面积雨水花园

3. Some building drainage pipes have been planted in green space 部分建筑排水管已经植入绿地

长不佳

3. Highland hard to keep water and let them fully infiltrated 大部分为平地以及高燥绿地无法收集路面雨水

runoff calculation 径流计算

STATUS QUO-areal 地块1现状　UBC Holistic Living Roof & LID Calculater LIBC 生态屋顶与低影响开发计算器

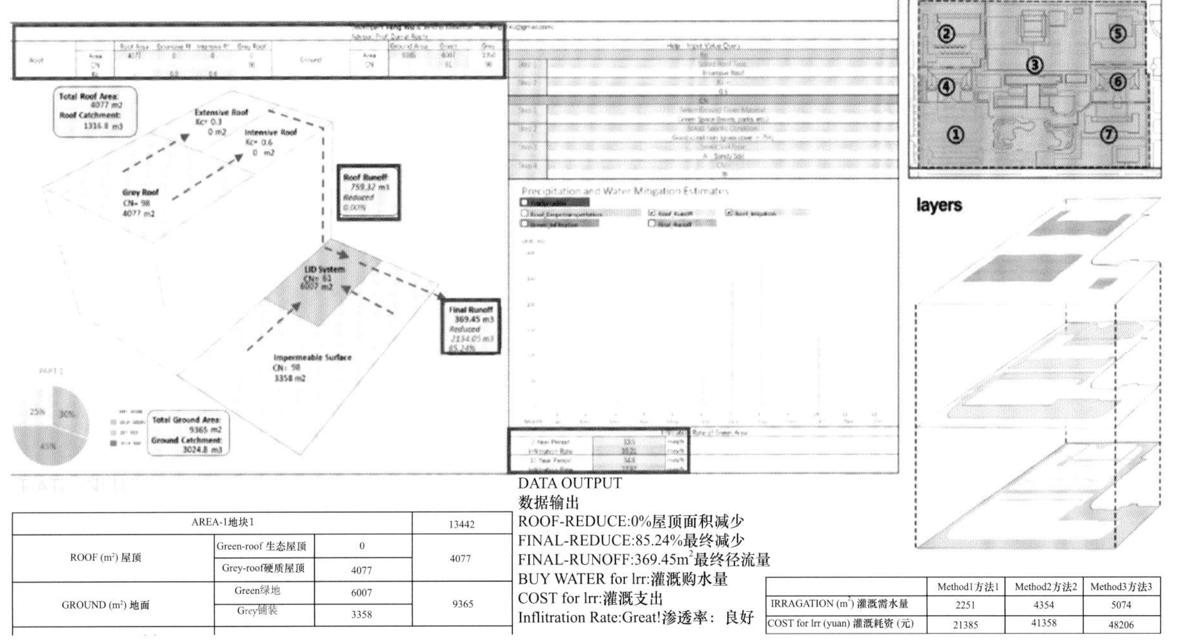

AREA-1地块1			13442
ROOF (m^2) 屋顶	Green-roof 生态屋顶	0	4077
	Grey-roof硬质屋顶	4077	
GROUND (m^2) 地面	Green绿地	6007	9365
	Grey铺装	3358	

	Method1方法1	Method2方法2	Method3方法3
IRRAGATION (m^2) 灌溉需水量	2251	4354	5074
COST for lrr (yuan) 灌溉耗资 (元)	21385	41358	48206

图 88　Site 1 status assessment
地块一现状评估

Use 6007m^2 green space to let 85% runoff infiltrated to the ground and use a tank to storage the rest 15% runoff for future irrigation.

利用 6007m^2 的绿地，让 85%的径流渗透到地下，并用水箱储存剩余的 15%径流，以备将来灌溉使用。

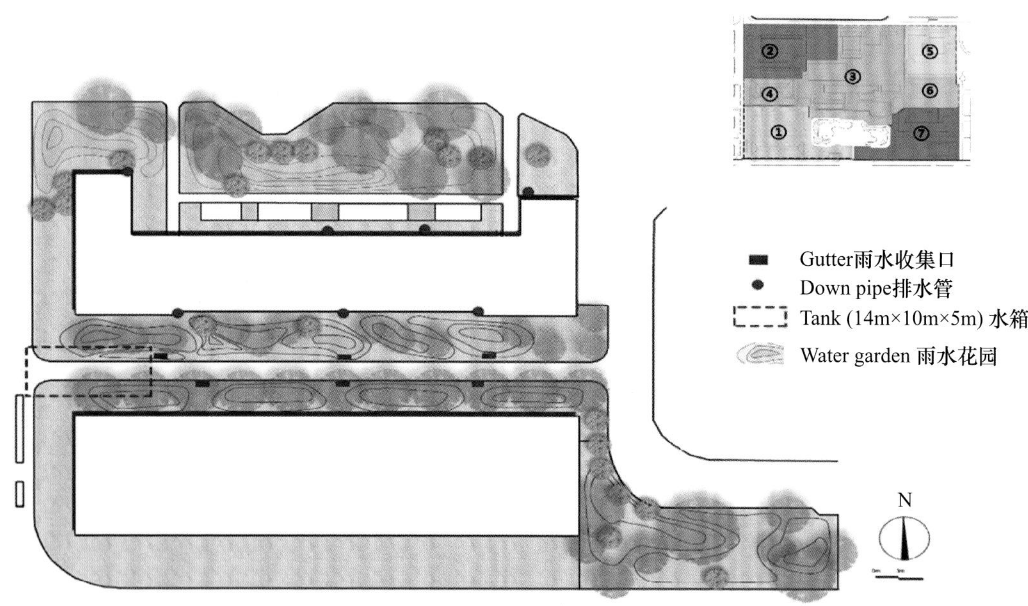

图 89　Site 1 plan
地块一平面图

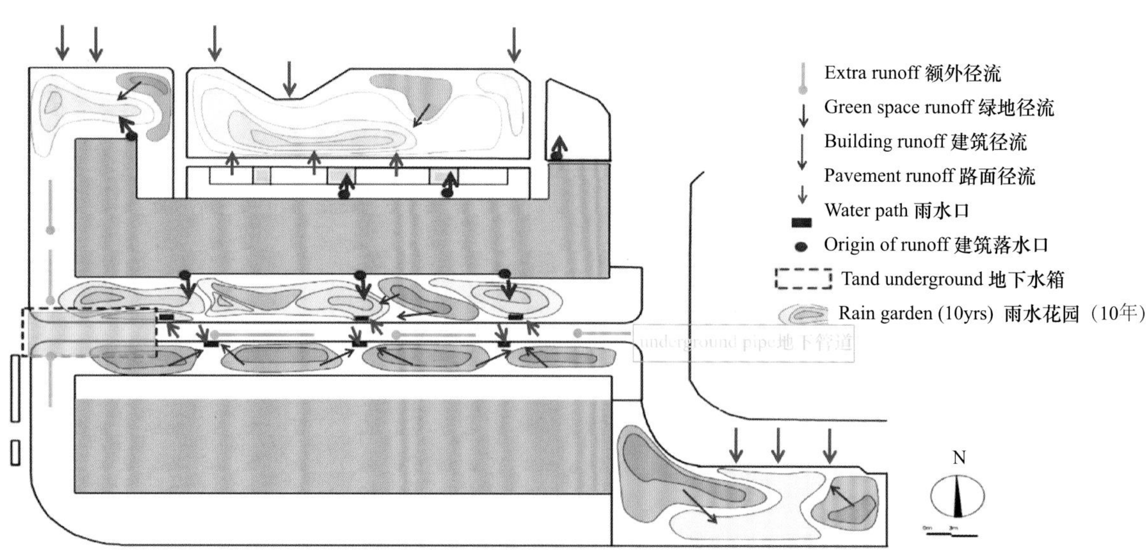

图 90　Site 1 design analysis
地块一设计分析图

乔木灌木保留 Preserved arbors

Juniperus chinensis 圆柏
Syringa reticulata 暴马丁香
Cerasus serrulata 日本晚樱
Pinus tabuliformis Carrière 油松
Crataegus pinnatifida Bunge 山楂
Cephalotaxus sinensis 粗榧
Platanus occidentalis 美桐
Pinus bangeana Zucc 白皮松
Lonicera maackii(Rupr.)Maxim 金银木

现状草本地被 Present herbs

Verbena officinalis L 马鞭草
Ophiopogon japonicus 麦冬
Hosta plantaginea 玉簪
Bambusoideae 竹子

Preserve the existing arbors and part Of the shrubs,and add more herbs and arbors resisting to water.
保留现有乔木和部分灌木，增加耐水乔木和草本植物

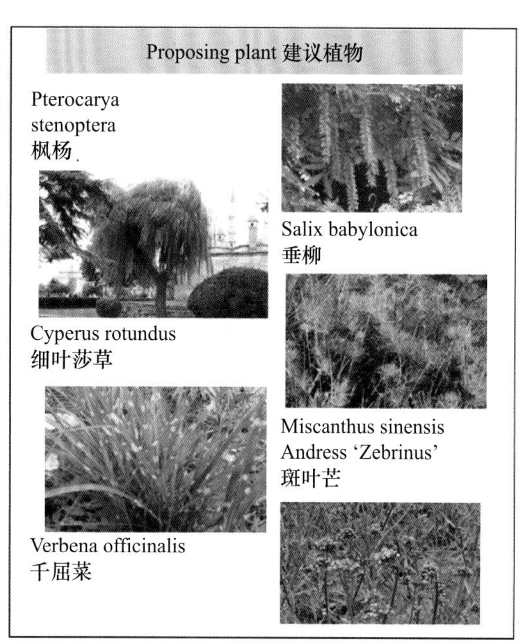

图 91 Site 1 plants arrangement
地块一植物配置

3.2.2 Site 2 地块二设计

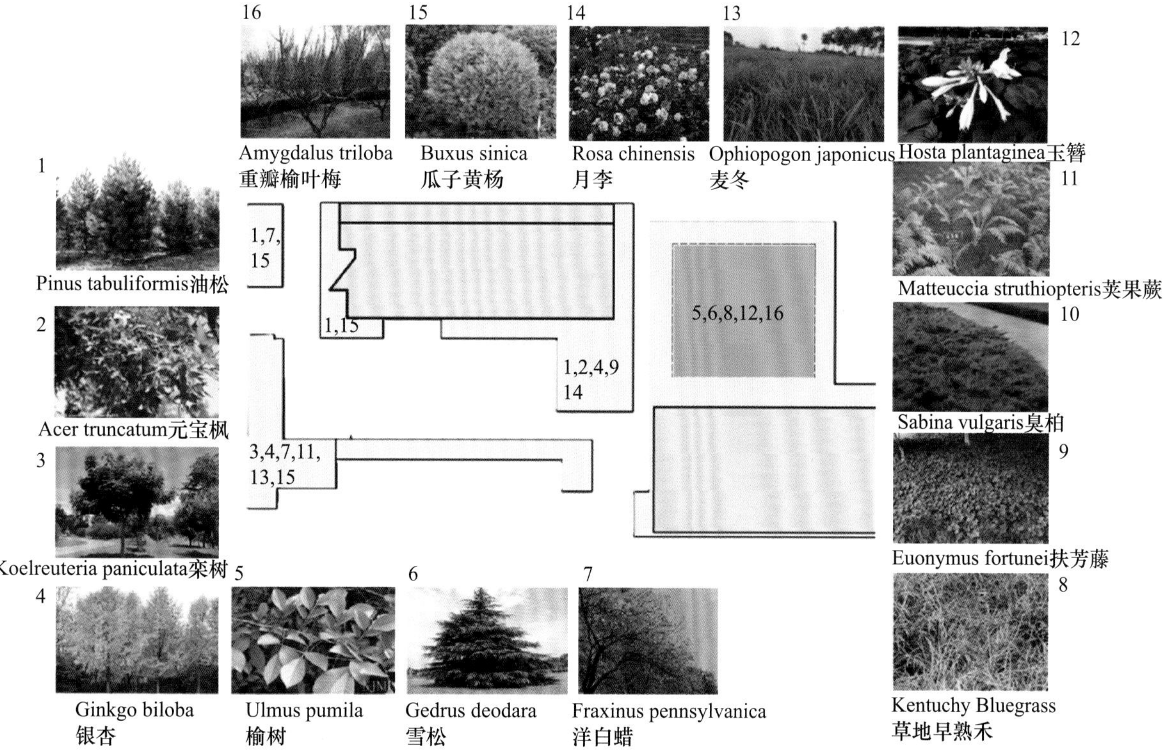

图 92 Site 2 current plants
地块二植物现状

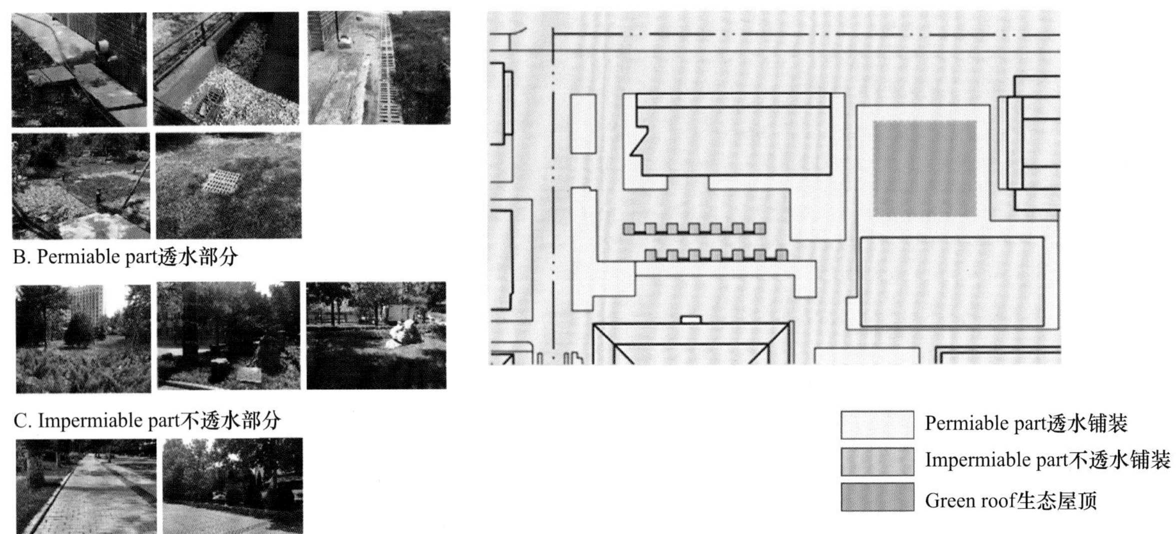

图 93 Site 2 current situation
地块二现状分析

Advantage 优点：

1. Green space rate is higher. 绿地率较高
2. There is a height difference. 场地有高差

Disadvantage 缺点：

1. Plant species are single and lack of sense of hierarchy. 植物种类单一，缺乏层次感
2. The pavement is impermeable. 路面不透水

表 22 GREEN SPACE TYPE
绿地种类

TYPE类型	EVALUATE评价	PHOTO 照片
Roof garden 屋顶花园	Rich in planting and surrounding catchment 植被丰富，集水面积大	
High in altitude and dry lawn 高于路面的草地	Road waner can not be imported. 路面雨水无法进入	
Planting pool 种植池	Water can not overflow. 雨水无法溢流	
Lawn 草地	Water can not overflow. 雨水无法溢流	

STATUS QUO-area2 地块2现状

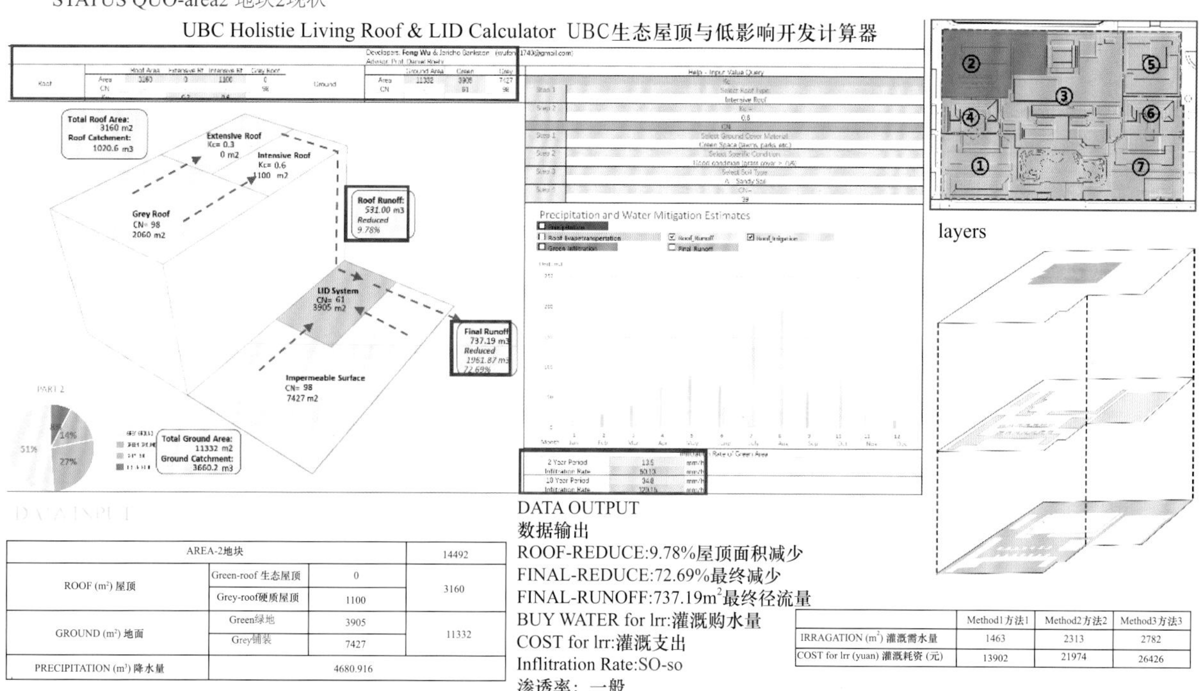

AREA-2地块			14492
ROOF (m²) 屋顶	Green-roof 生态屋顶	0	3160
	Grey-roof硬质屋顶	1100	
GROUND (m²) 地面	Green绿地	3905	11332
	Grey铺装	7427	
PRECIPITATION (m³) 降水量	4680.916		

	Method1方法1	Method2方法2	Method3方法3
IRRAGATION (m³) 灌溉需水量	1463	2313	2782
COST for lrr (yuan) 灌溉耗资(元)	13902	21974	26426

图 94 Site 2 status assessment
地块二现状评估

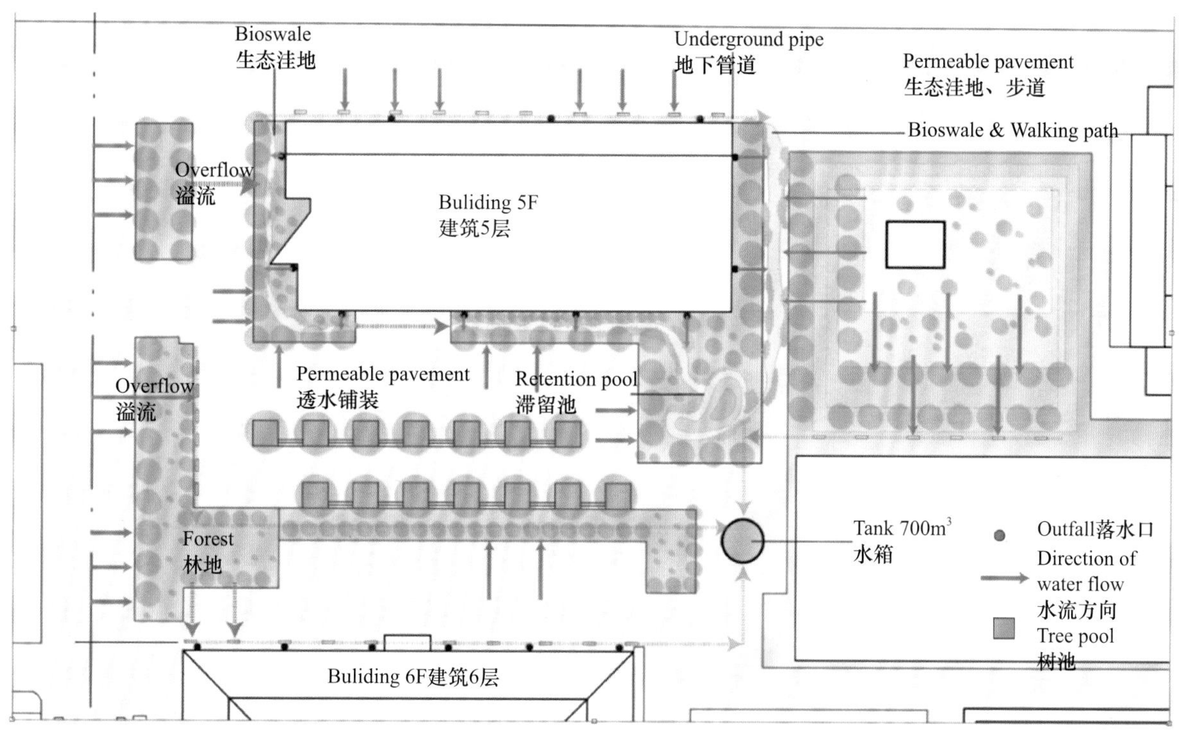

图 95 Site 2 plan
地块二平面图

EXEMPLES 示例

Retention pool滞留池

Bioswale & Walking Path
生态洼地、步道

SECTION 剖面图

Evaporation
蒸发

Downpipe
落水管

Buliding
建筑

Downpipe
落水管

Buliding
建筑

Rainwater garden
雨水花园

Overflow
溢流

Bioswales
生态洼地

Overflow
溢流

Permeable pavement
透水铺装

Permeable pavement
透水铺装

Infiltration
下渗

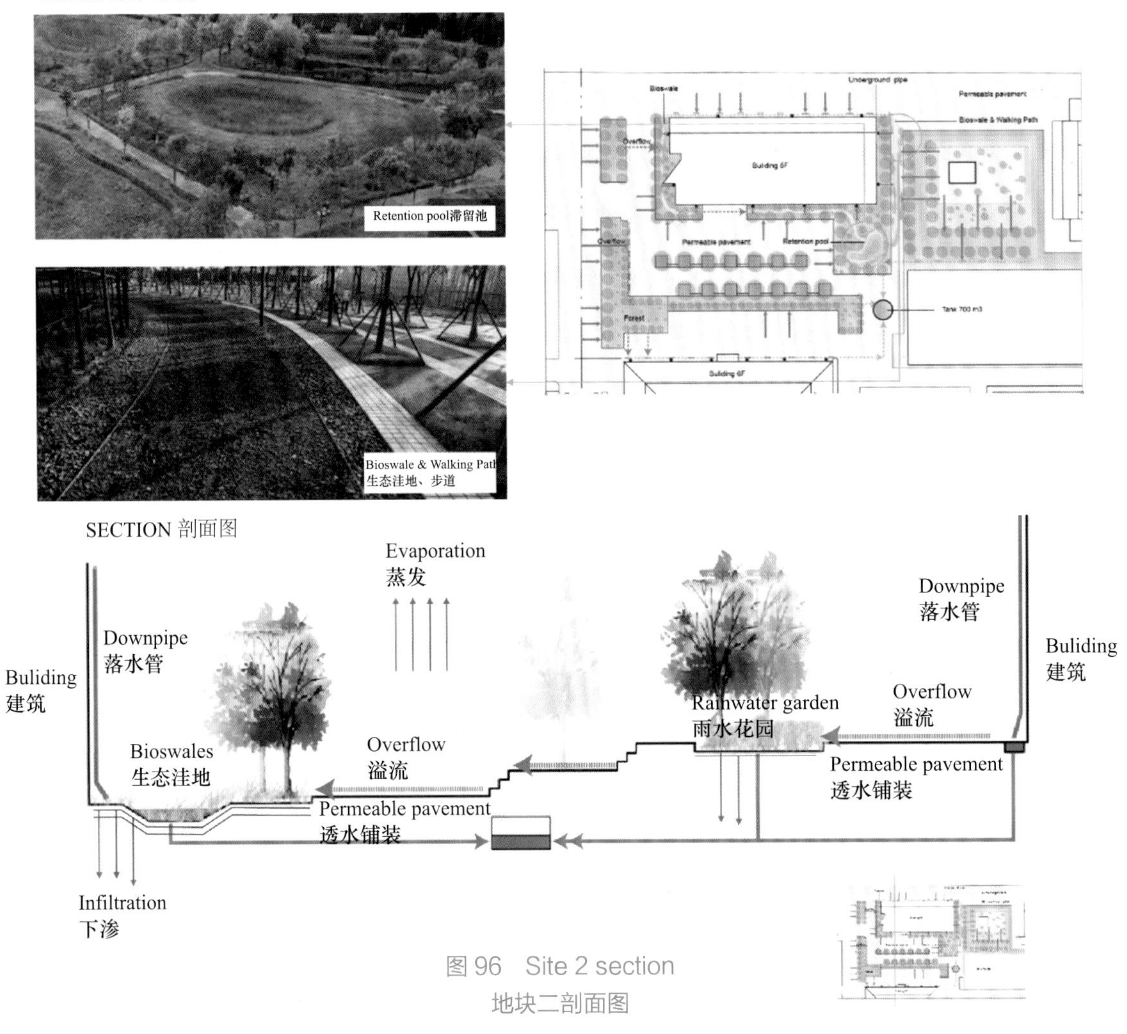

图 96　Site 2 section
地块二剖面图

EXEMPLES 示例

图 97　Transformation examples of site 2
地块二改造示例

BEFORE AND AFTER SITE COMPARISON 设计前后对比

BEFORE 设计前	**AFTER 设计后**
Final Runoff 最终径流量	**Final Runoff 最终径流量**
737.19m^3	352.13m^3
Reduced 减少	*Reduced 减少*
1961.87m^3	*2346.92m^3*
72.69%	*86.95%*

3.2.3 Site 3 地块三设计

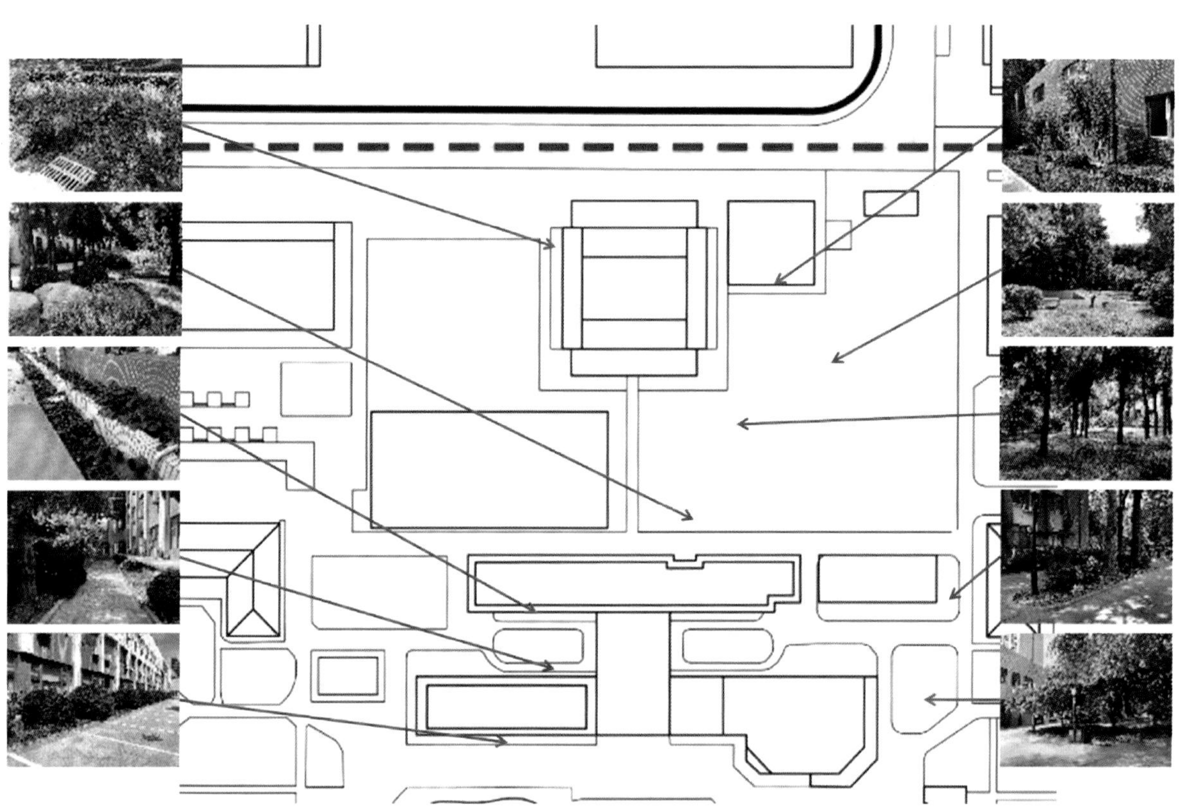

图 98 Site 3 Situation of green space 1
地块三现状绿地分析 1

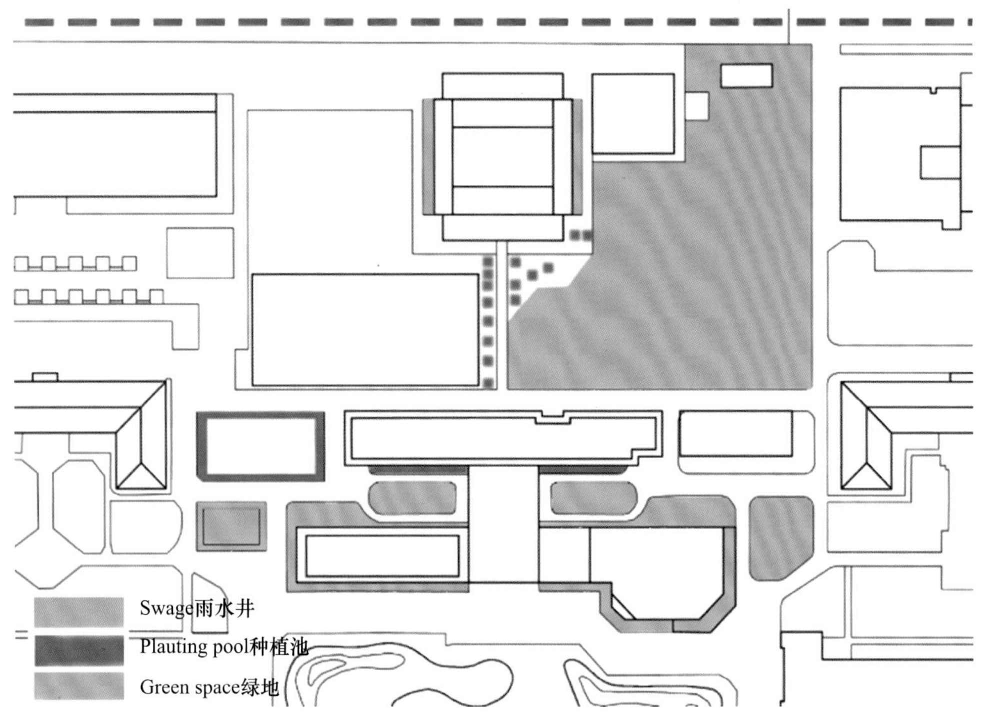

图 99 Site 3 Situation of green space 2
地块三现状绿地分析 2

Advantages 优点：

1.Green space rate is high. 绿地率较高

2.Good landscape effect. 景观效果好

Disadvantages 缺点：

1.The rain on building falls into the pipe network directly. 建筑落水直接进入管网

2.The pavement is impermeable. 路面不透水

表 21 Site 3 Current landscape assessment
地块三现状景观评价

Type种类	Evalutation评价	photo照片
Bioswale 洼地花境	Full infiltration, good landscape effect 充分下渗景观效果好	
Planting pool 种植池	Water cannot overflow 积水无法溢流	
Green space 绿地	Unable to accommodate surrounding runoff 无法容纳周围径流	

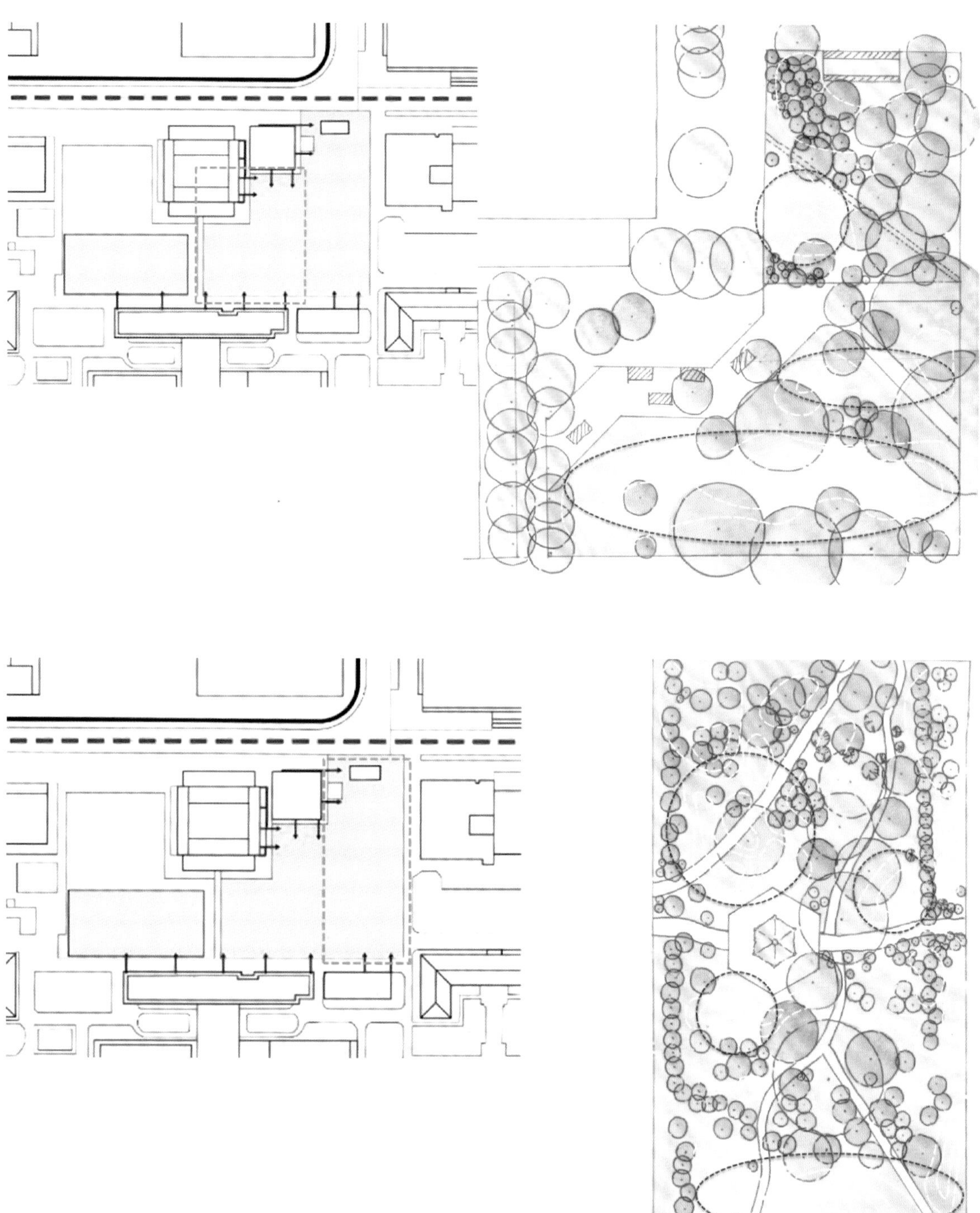

图 100　Site 3 Design plan
地块三设计平面

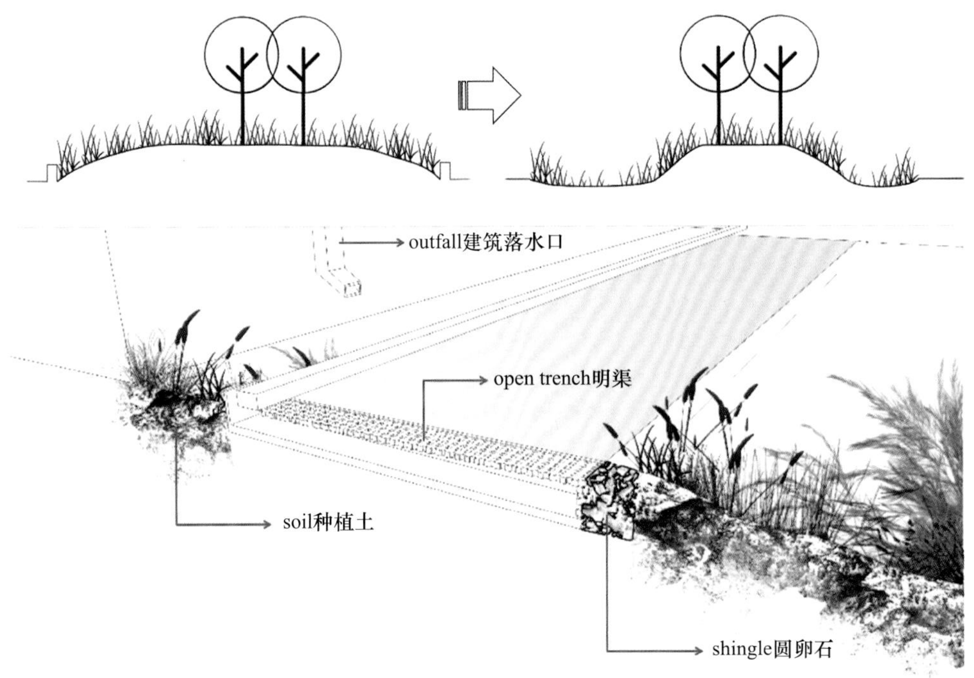

图 101　Site 3 Design details
地块三设计细节

3.2.4　Site 4 地块四设计

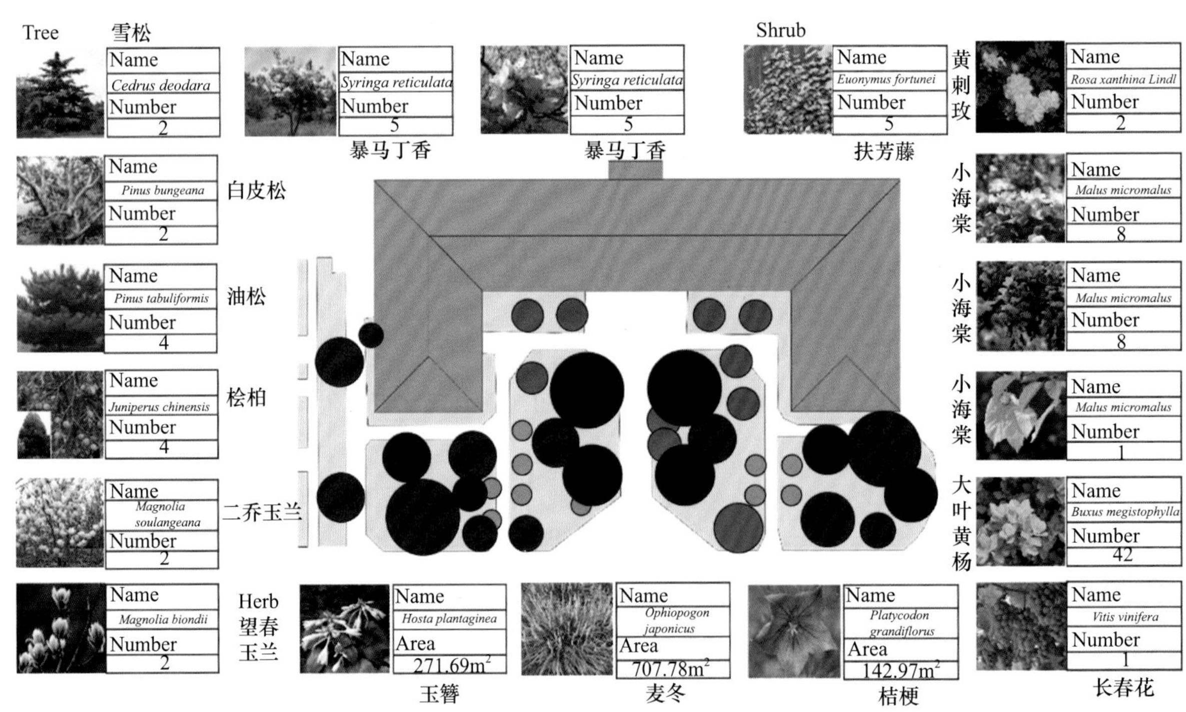

图 102　Site 4 Current plants
地块四植物现状

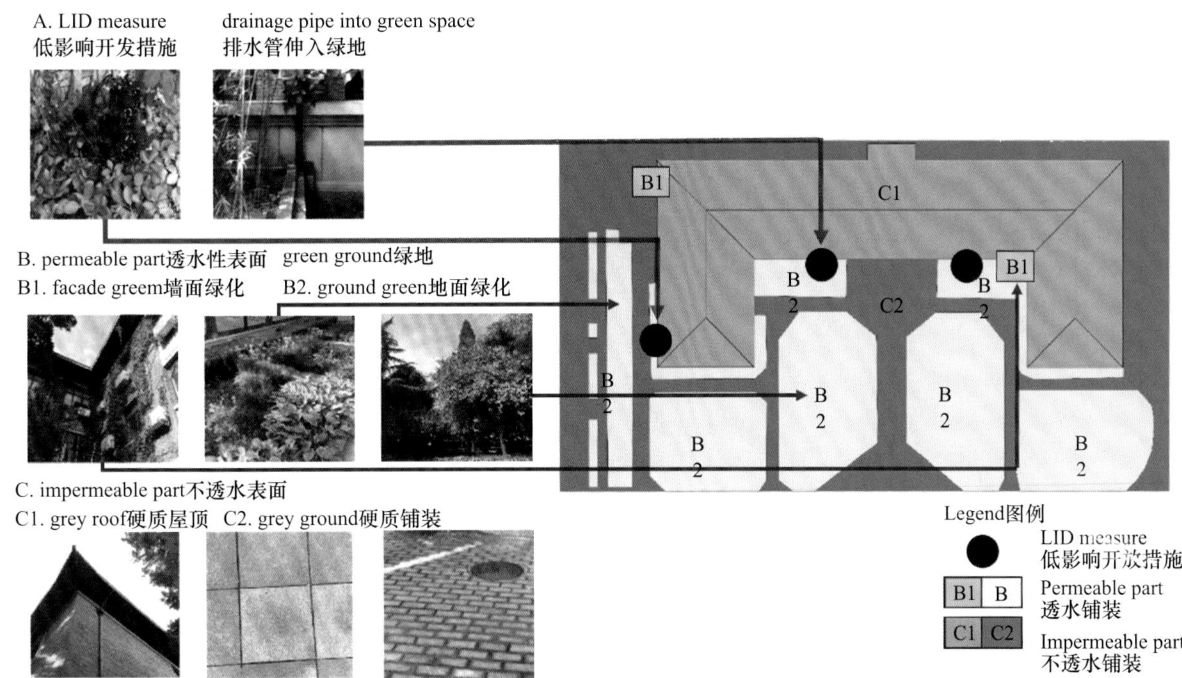

图 103　Site 4 Situation analysis
地块四现状分析

Advantages 优点：

1.The green space and the building are axisymmetric,with a beautiful configuration.

绿色空间和建筑是轴对称的，具有漂亮的配置。

2.Green walls have various viewing experience in four seasons.

绿墙在四季都有各种观赏体验。

Disadvantages 缺点：

1.There are no LID measures in the north side.

北侧没有低影响开发措施。

2.All pavings are impervious.

所有铺装都是不透水的。

表 21　Current landscape assessment of site 4
地块四现状景观评价

Type类型	Comment评价	Photo照片
Vertical Greening 立体绿化	Good landescape on the facade. 墙面景观效果好	
Planter 花池	Water can not overflow. 积水无法溢流	
Lawn 草坪	Road water can not be remitted. 无法吸纳路面积水	

Ground Type铺装类型

Too much impervious surface leads to
过多不透水铺装导致

· Sewer overflow水管溢流
· Flooding洪水
· Degradation of water quality水质下降

Flooding in July.七月雨水现状

Flooding area积水区域

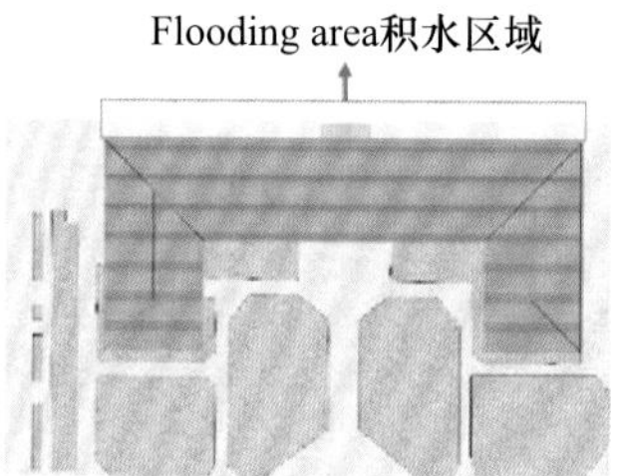

My design goal is
To intergrate:
设计目标–转化

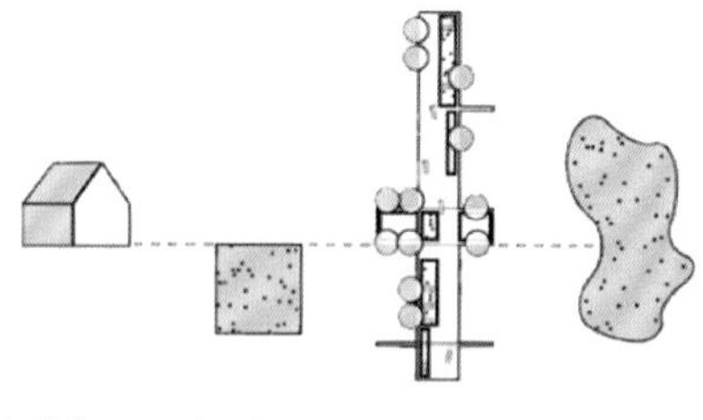

building 建筑 | landscape 景观 | street 街道 | open space 开放空间

First Step第一步

· Runoff reduction减少径流

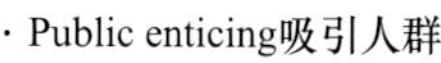

Long Term Benefit长期优势

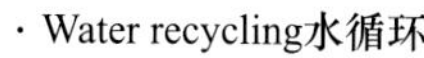

· Viewing experience美化景观

· Public enticing吸引人群

· Water recycling水循环

图 104 Site 4 Design ideas
地块四设计思路

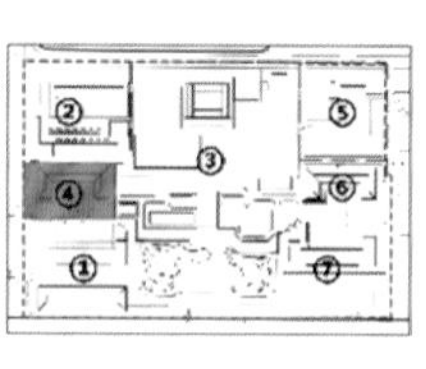

Legend图例

- Downpipe排水口
- Open drain排水沟
- Rain garden雨水花园
- Bio swale生态洼地
- Permeable paving 透水铺装
- Impermeable paving 不透水铺装
- Vertical green 垂直绿化

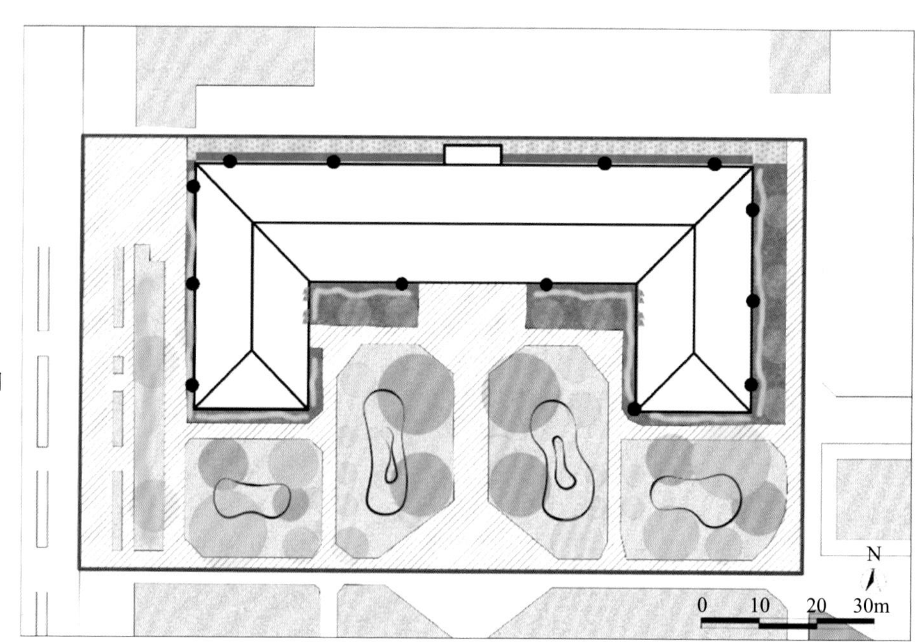

图 105 Site 4 Design plan
地块四设计平面

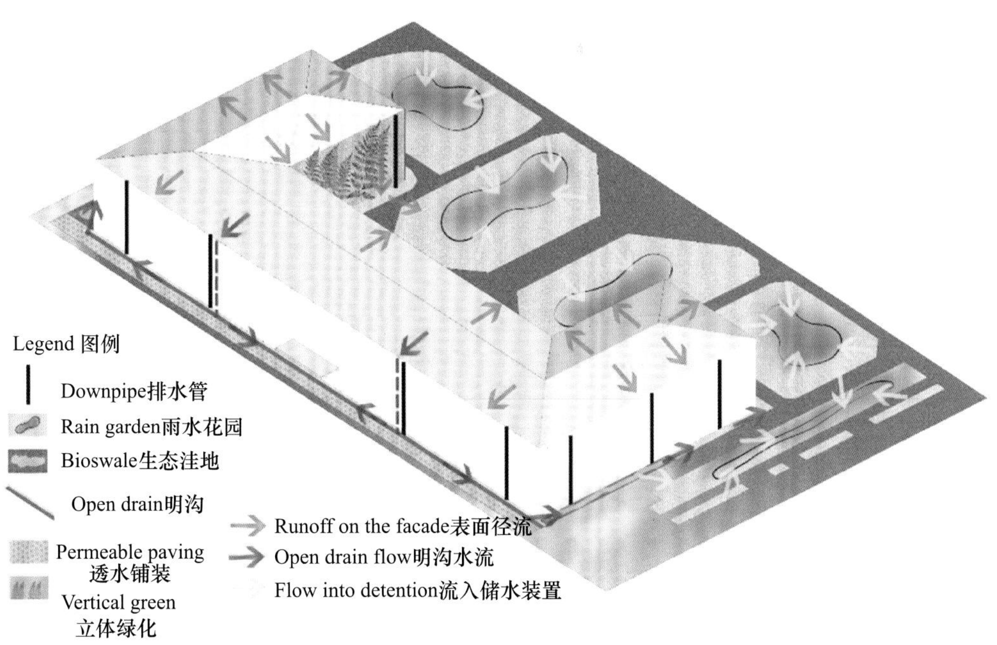

图 106　Site 4 design components

地块四设计元素

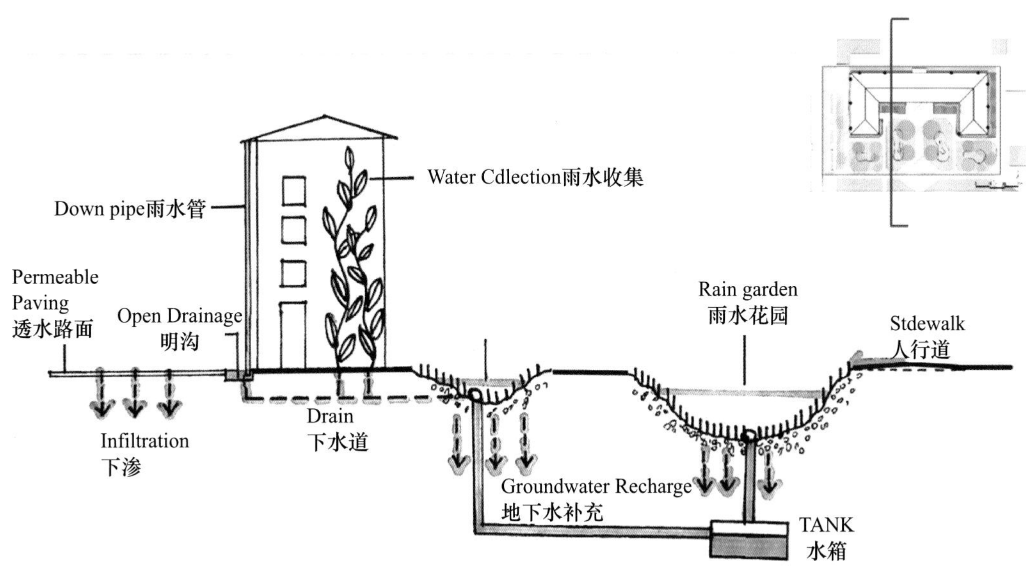

图 107　Site 4 site holistic system section

地块四系统剖面

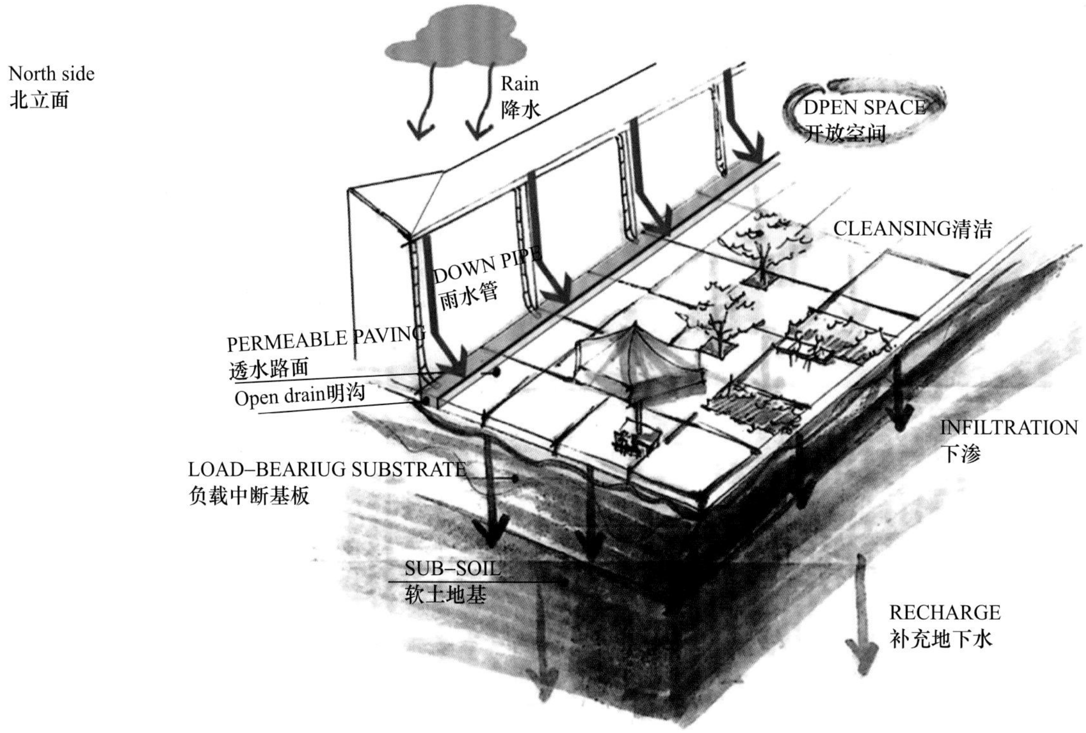

North side
北立面

Rain
降水

DPEN SPACE
开放空间

CLEANSING清洁

DOWN PIPE
雨水管

PERMEABLE PAVING
透水路面

Open drain明沟

INFILTRATION
下渗

LOAD–BEARIUG SUBSTRATE
负载中断基板

SUB–SOIL
软土地基

RECHARGE
补充地下水

图 108 Rain path
雨水路径

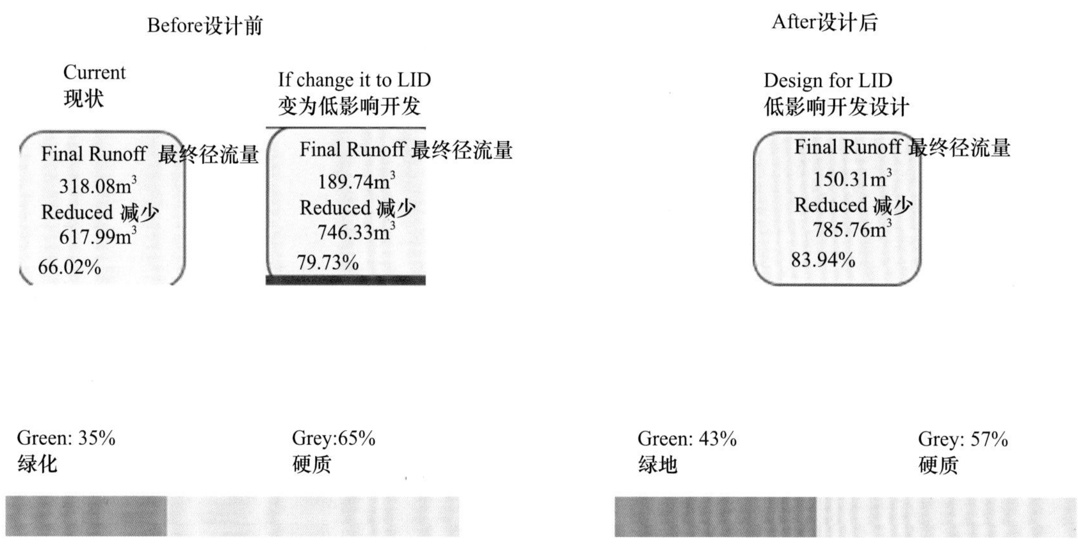

图 109 Comparison of effect before and after design of site 4
地块四设计前后效果比较

3.2.5 Renewal Design of Main Building 主楼更新设计

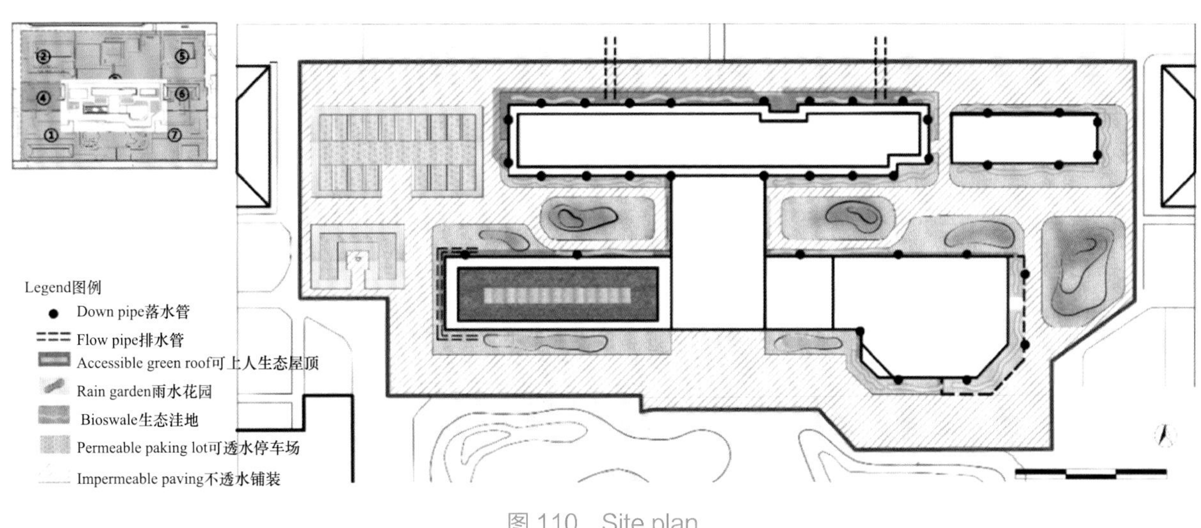

图 110 Site plan
场地平面图

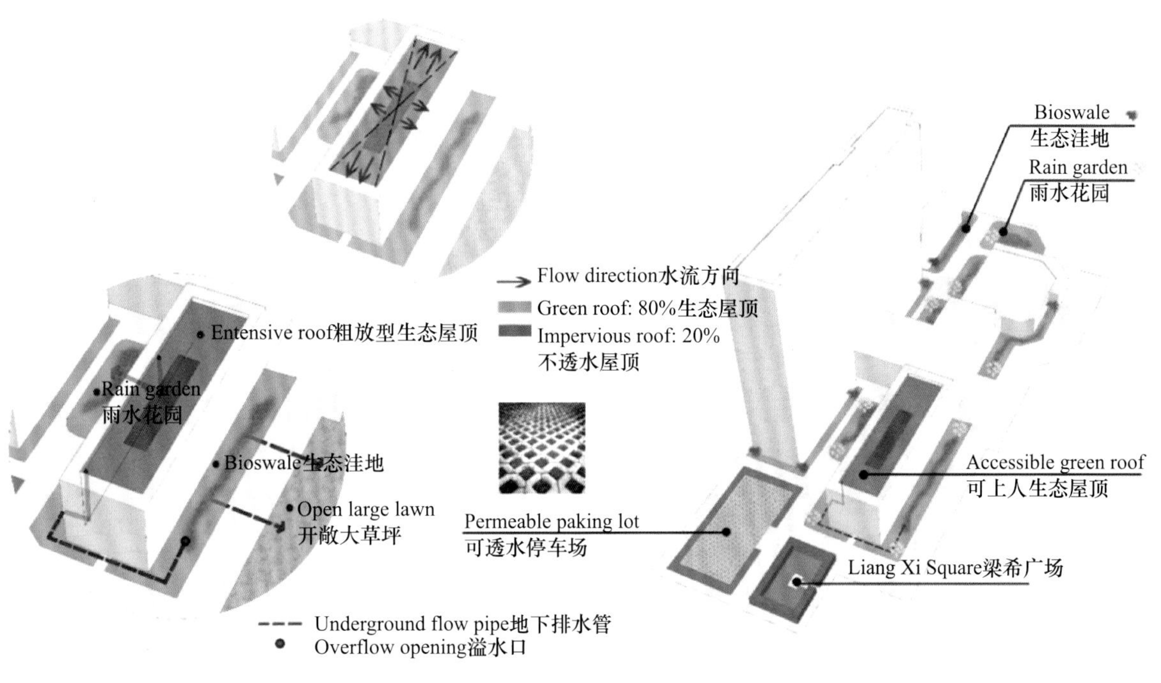

图 111 Design components
设计内容

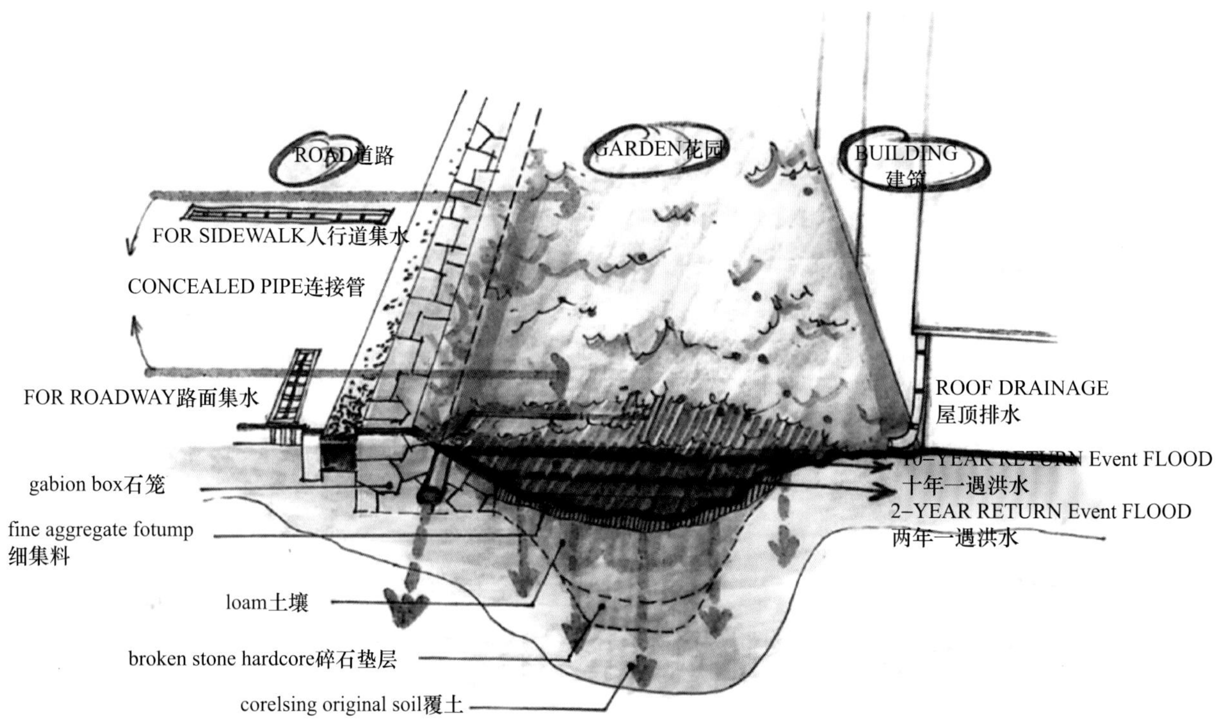

图 112　Rain garden section
雨水花园剖面图

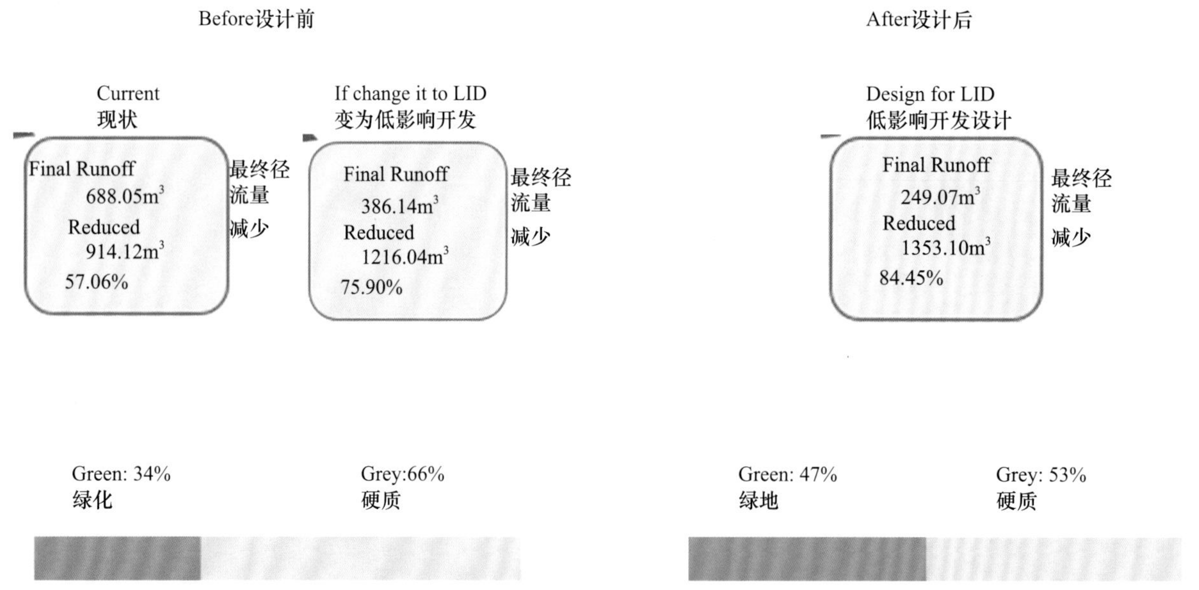

图 113　Effect comparison of main building area before and after design
主楼区域设计前后效果比较

3.2.6 Site 5 地块五设计

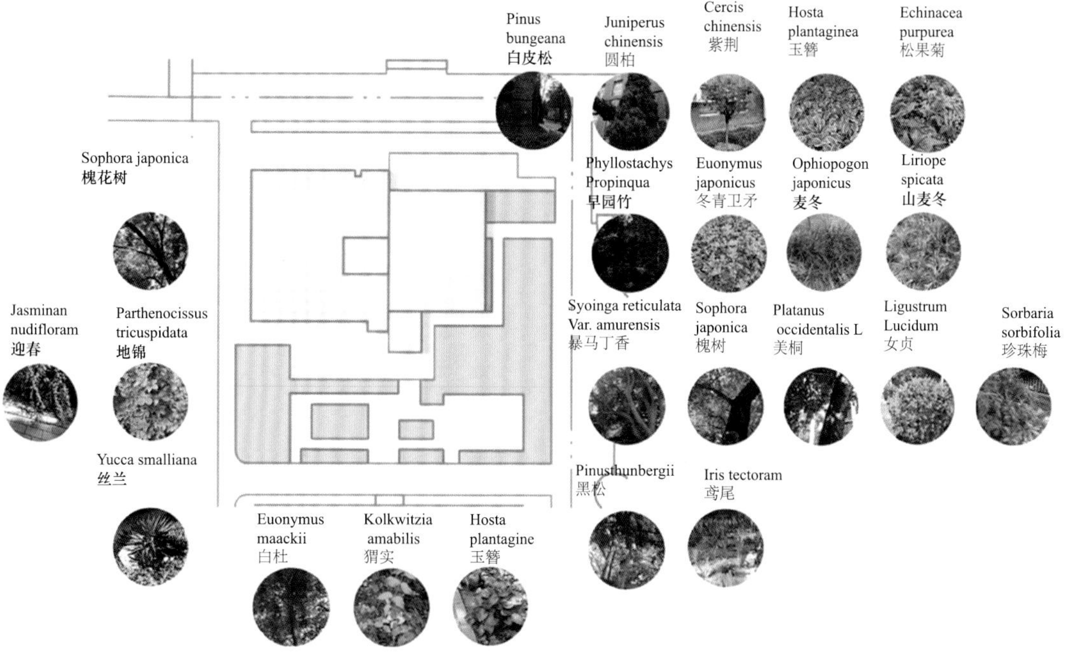

图 114 Site 5 Current plants
地块五植物现状

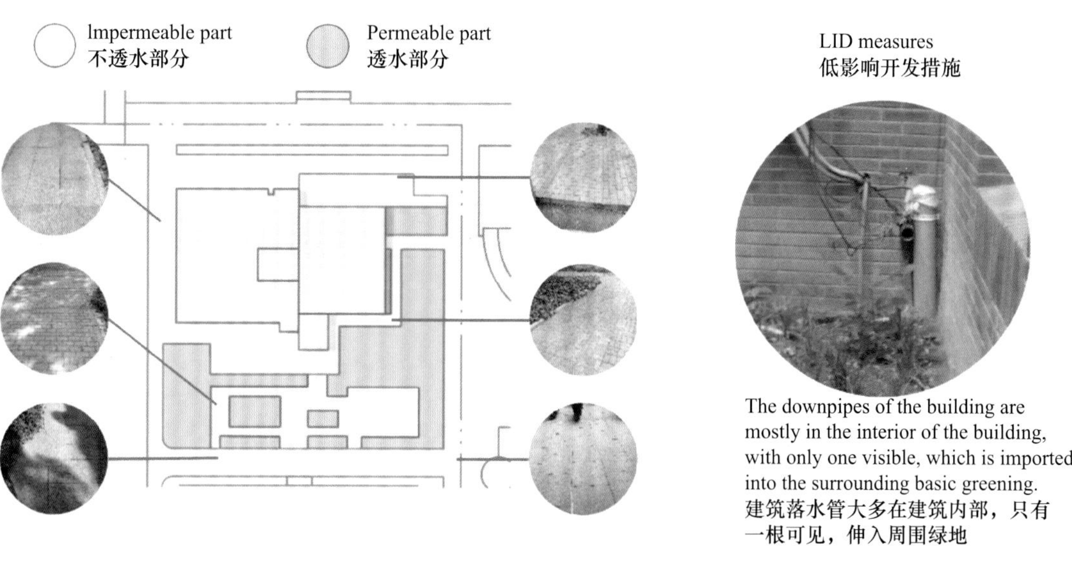

图 115 Site 5 Situation analysis
地块五现状分析

Site advatages 场地优势

1.High greening rate and abundant plant species in the site. 绿地率高，植物种类丰富

2.Most of the building drainage pipes are hidden inside the building, which is more beautiful. 多数建筑物排水管在建筑内，美观

Site defects 场地劣势

1.All paving are impermeable paving. 均为不透水铺装

2.There are kerbs on the edge of green space, which is not conducive to drainage. 路缘石不利于积水汇入

3.Lack of basic greening around the building. 建筑周边绿地缺乏

4.Part of the construction downpipe runoff is directly drained to the impervious pavement. 部分排水管直接将水排至不透水铺装

Type 绿地类别	Evalutation 评价	Photo 照片
High and dry green space 高燥绿地	High in the ceuter, not conducive to rain water access 中心高，不利于雨水汇入	
Planting pool 种植池	a escessive difference from road in height 与道路高差过大且未作其他处理，不利于积水汇入	
lawn 草坪	Curb sfones are not conducive to road wafer inflow 路缘石不利于道路积水汇入	

图 116　Site 5 Situation analysis
地块五现状分析

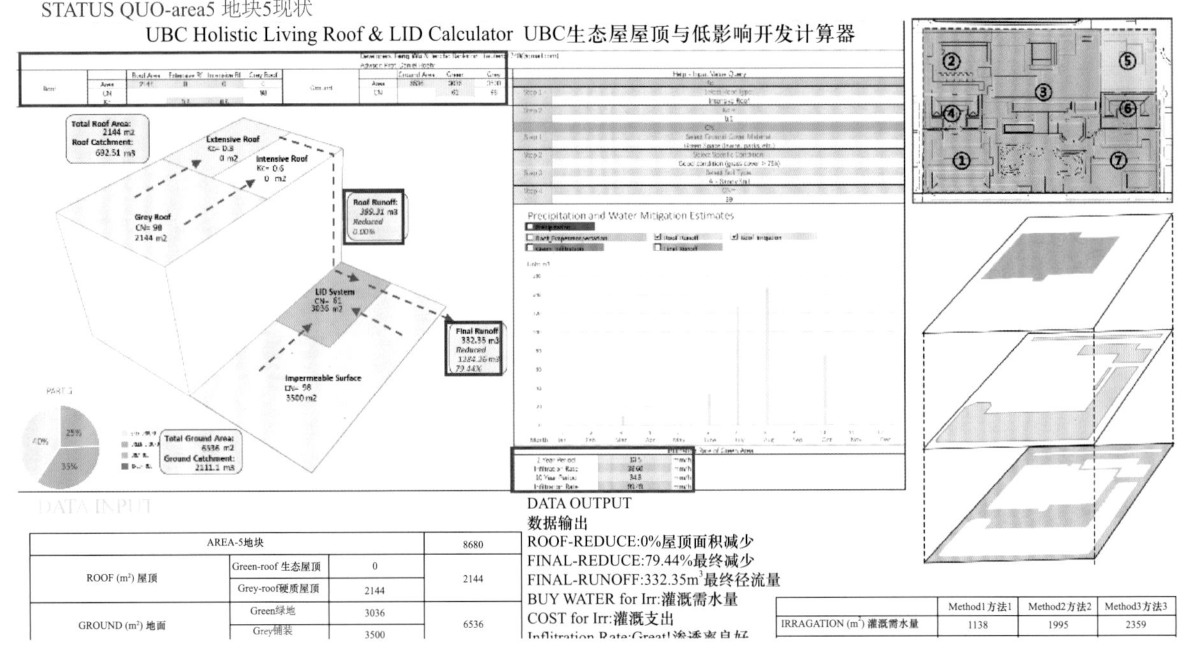

图 117　Site 5 Status assessment
地块五现状评估

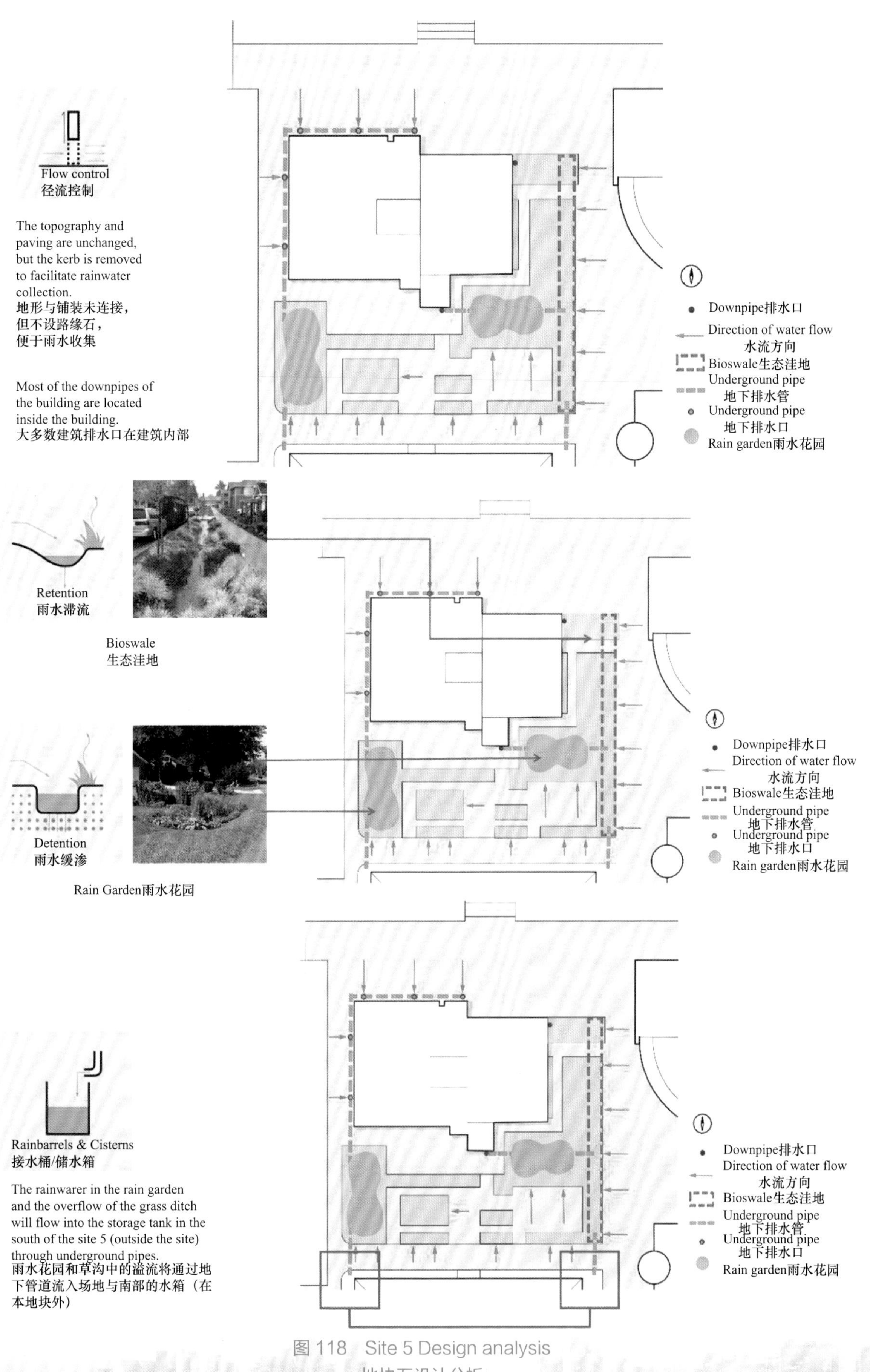

图 118　Site 5 Design analysis
地块五设计分析

3.2.7 Site 6 地块六设计

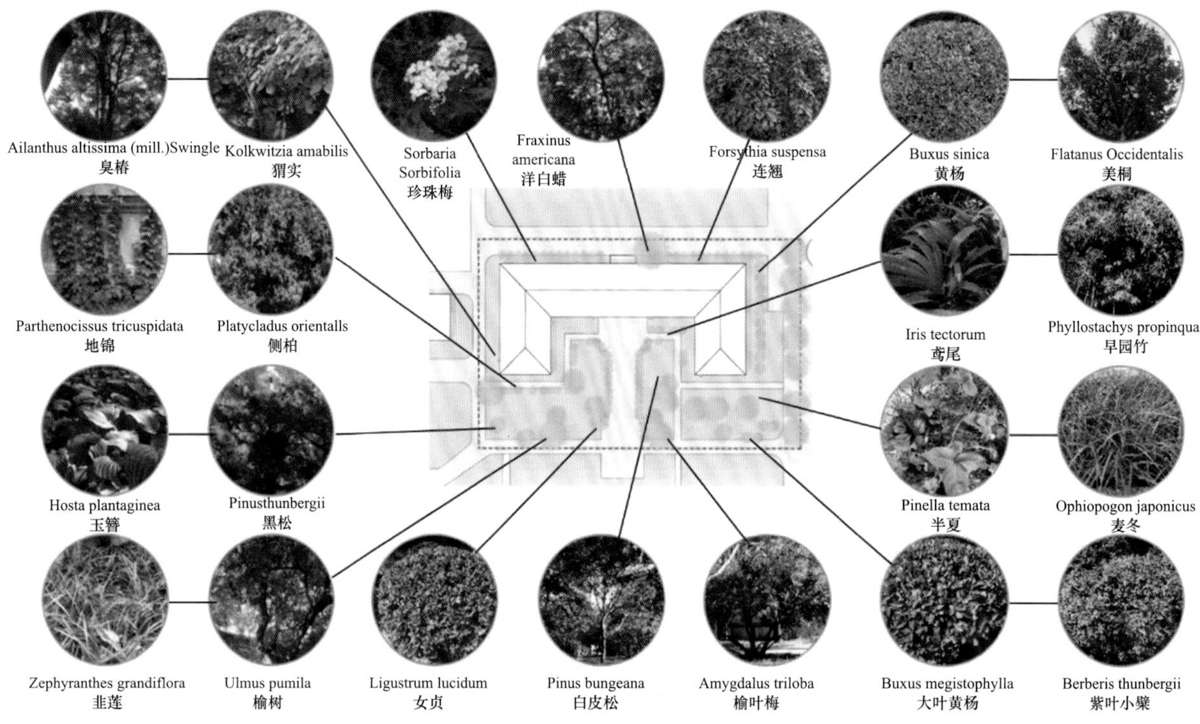

图 119　Site 6 Current plants
地块六植物现状

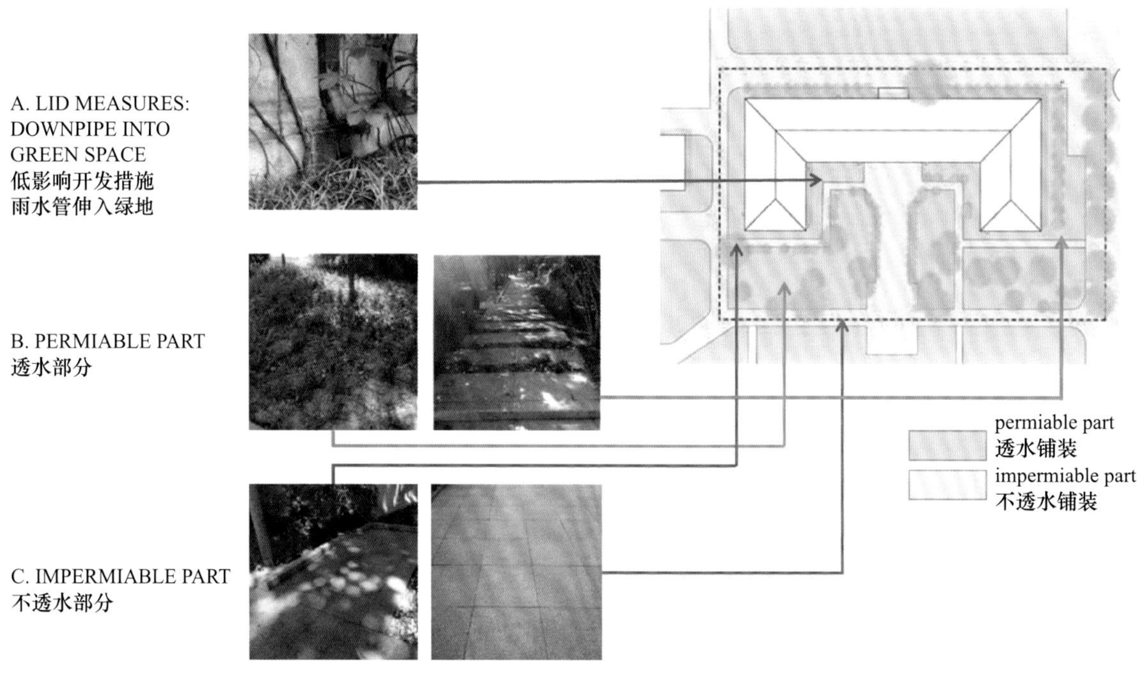

图 120　Site 6 Situation analysis
地块六现状分析

Advantages 优势：

1.Green rate is higher.

绿地率较高

2.All of the drainage pipes in the building are fed into the green space, making it easy for water to leak.

所有的建筑排水管都导进了绿地中，易于渗水

Disadvantages 劣势：

1.The plant species is single and lacks layering.

植物种类单一，缺少层次感

2.The carvings are made of marble tiles.

铺装均为大理石砖面

表 22　Current landscape assessment
地块六现状景观评价

variety 种类	analysis 分析	picture 照片
High and dry green space 高燥绿地	Although there is a slope, the shade of surrounding trees is not obvious in the periphery,and the form is not conducive to drainage. 虽然有坡度，但是由于周围乔木的遮挡在外围看不明显，且中心高四周低的形式不利于排水。	
cultivation pool 种植池	There are street teeth around to separate the planting pond from the road, and the water on the road can not be discharged into the planting pond. 周围有马路牙子将种植池与道路分开，道路上的积水无法排入种植池。	
lawn 草坪	The lawn is more concentrated, and there is also the situation that water cannot be discharged. 草坪分布较集中，同样存在积水无法排入的情况。	

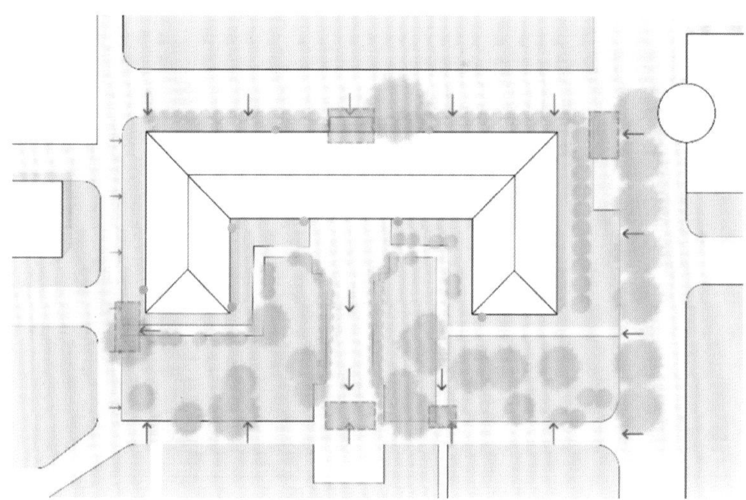

Down pipe落水管

Slope direction坡向

Water catchment area积水收集区域

图 121　Site 6 Rainwater status
地块六雨水现状

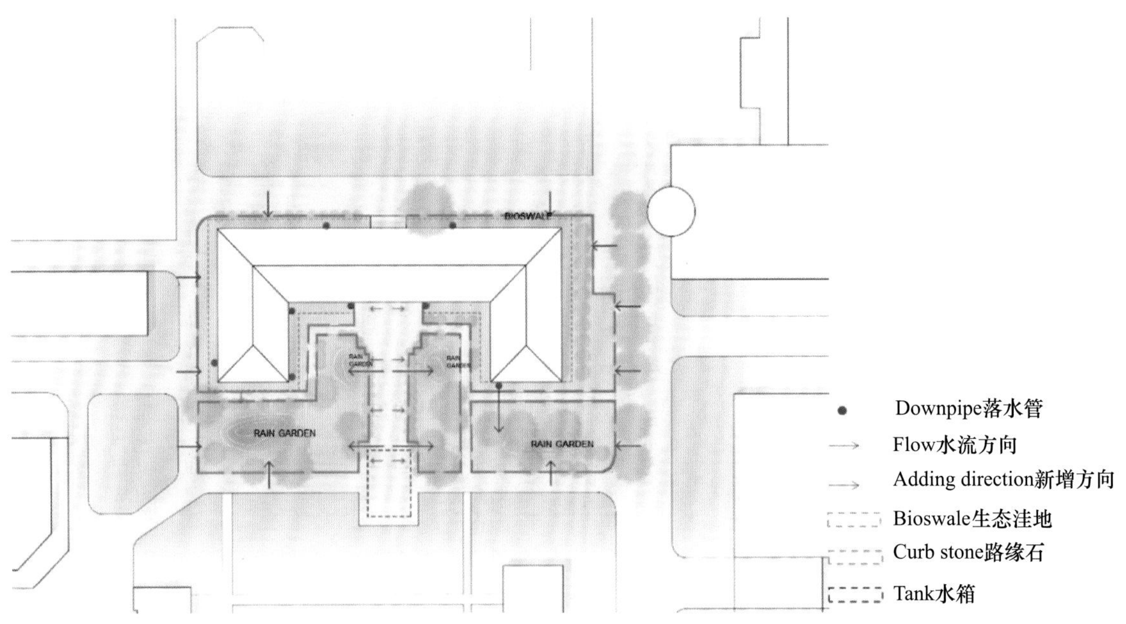

图 122　Site 6 Modified plan
地块六改进平面

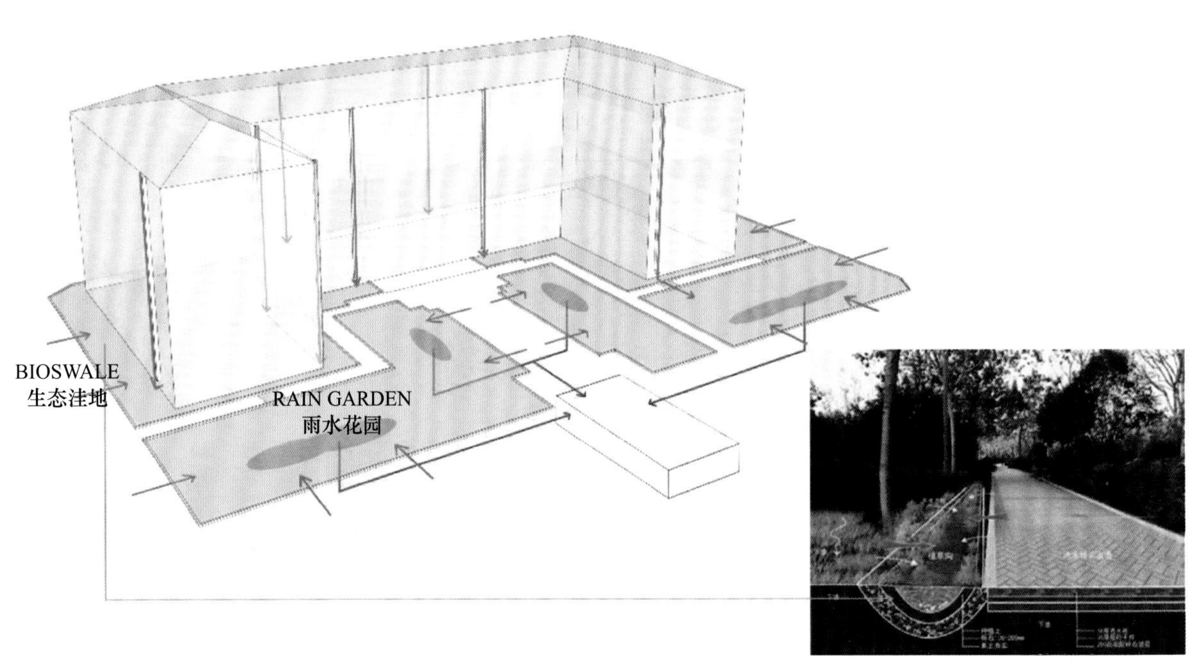

图 123　Site 6 Modified perspective
地块六改进办法

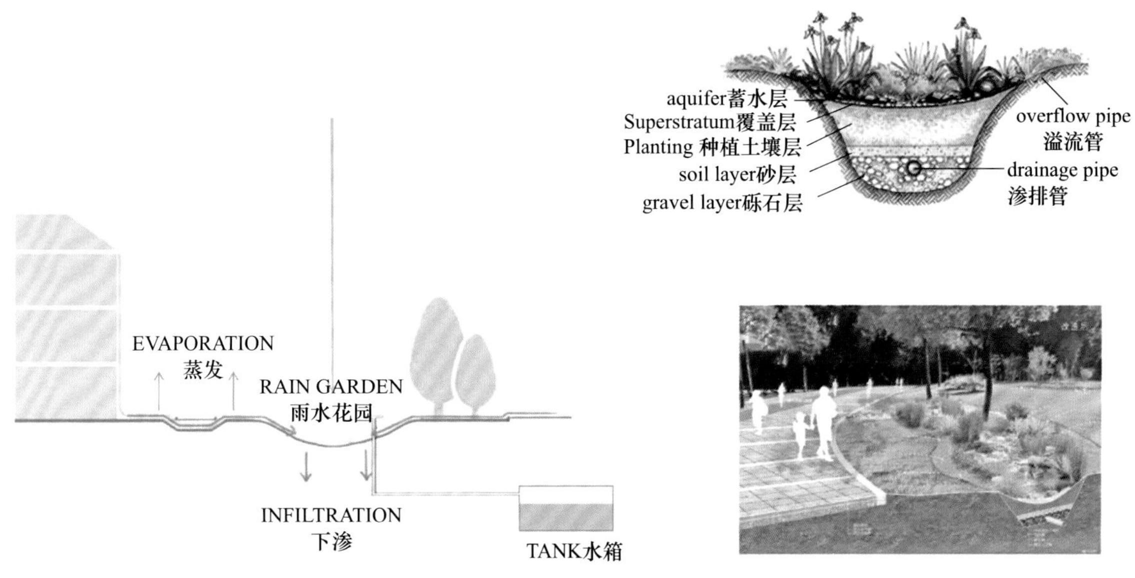

图 124 Site 6 Section
地块六剖面

3.2.8 Site 7 地块七设计

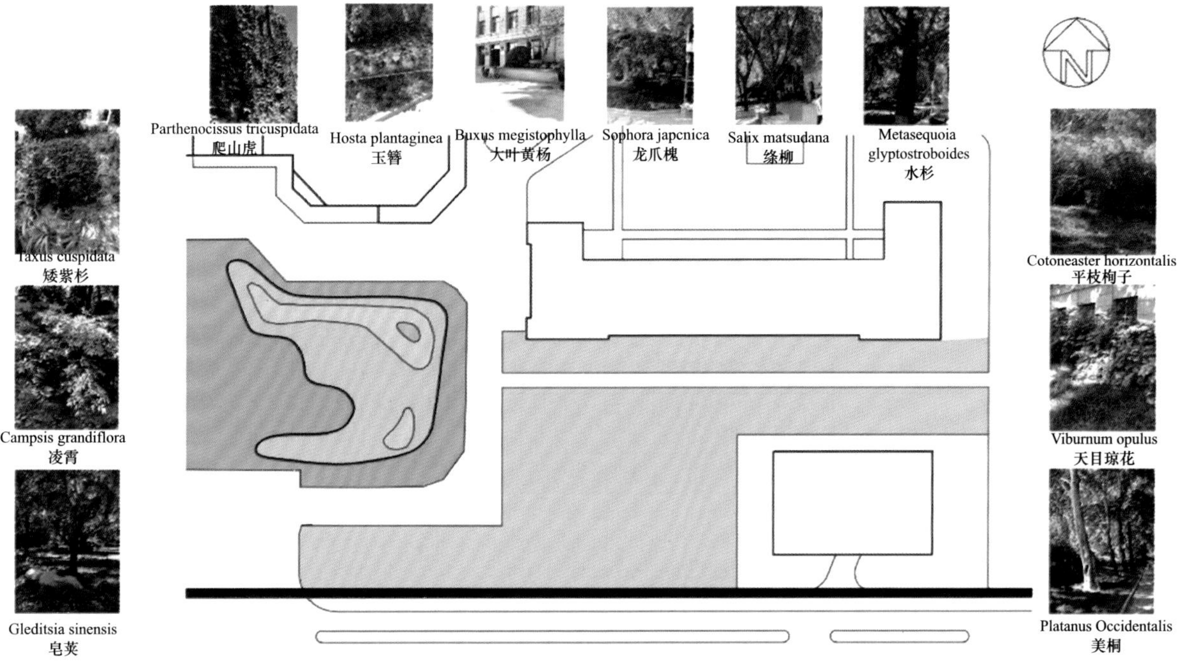

图 125 Site 7 Current plants
地块七植物现状

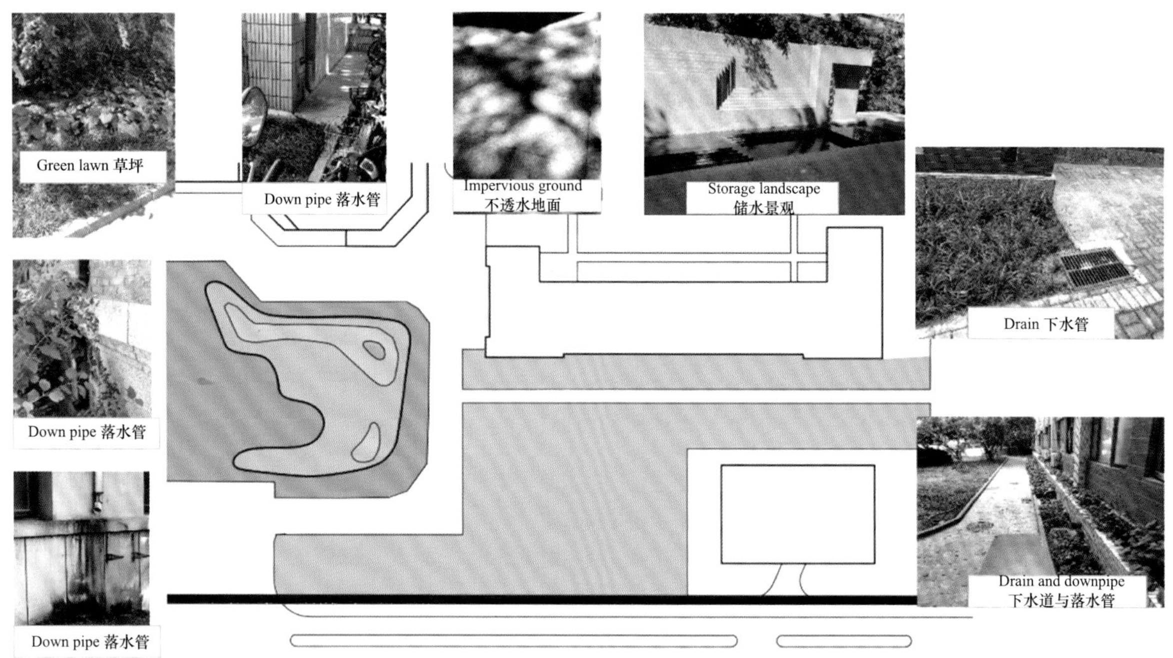

图 126 Site 7 Situation analysis
地块七现状分析

Advantages 优点：

1.Relatively permeable ground

透水地面相对较多

2.Rich plant species

植物种类较丰富，多样性较强

Disadvantages 缺点：

1.Light conditions are general, affecting plant growth.

光照条件一般，影响植物生长

2.Building downpipe is not associated with green space.

建筑落水管与绿地没有关联

3.The location of the rainwater well is high, and easy to accumulate water.

雨水井设置位置地势较高，易积水

表 23 Current landscape assessment
地块七现状景观评价

variety 种类	analysis 分析	picture 照片
1. High and dry green space 高燥绿地	The center is high and low around. Trees, shiubs and grass layer clearly and grow well. But plants have limited help in drainage from surrounding sites, mainly to solve their own rainwater problems and sometimes need additional irrigation. 中心高四周低，乔灌草分层较为明显，植物长势较好，但对于周边场地的排水帮助有限，主要解决自身雨水问题，有时还需要额外灌溉。	
2. cultivation pool 种植池	Mainly located near the entrance of the building The absorption capacity of rainwater is limited by area limitation. Can only accommodate part of the building roof falling water. 主要位于建筑入口附近，受面积限制对于雨洪吸收能力有限，只能容纳部分建筑屋顶落水。	
3. lawn 草坪	Lawn has drainage conditions, but because of the roadside barriers, the rainwater from the road cannot be led to planting pouds. Rainwater can be led to the lawn from the downpipe of the buildings, but the amount is limited. 有排水条件，但由于路缘石的阻碍，路上的雨水无法引入种植池中，部分建筑落水口处雨水可以引入，但数量有限。	

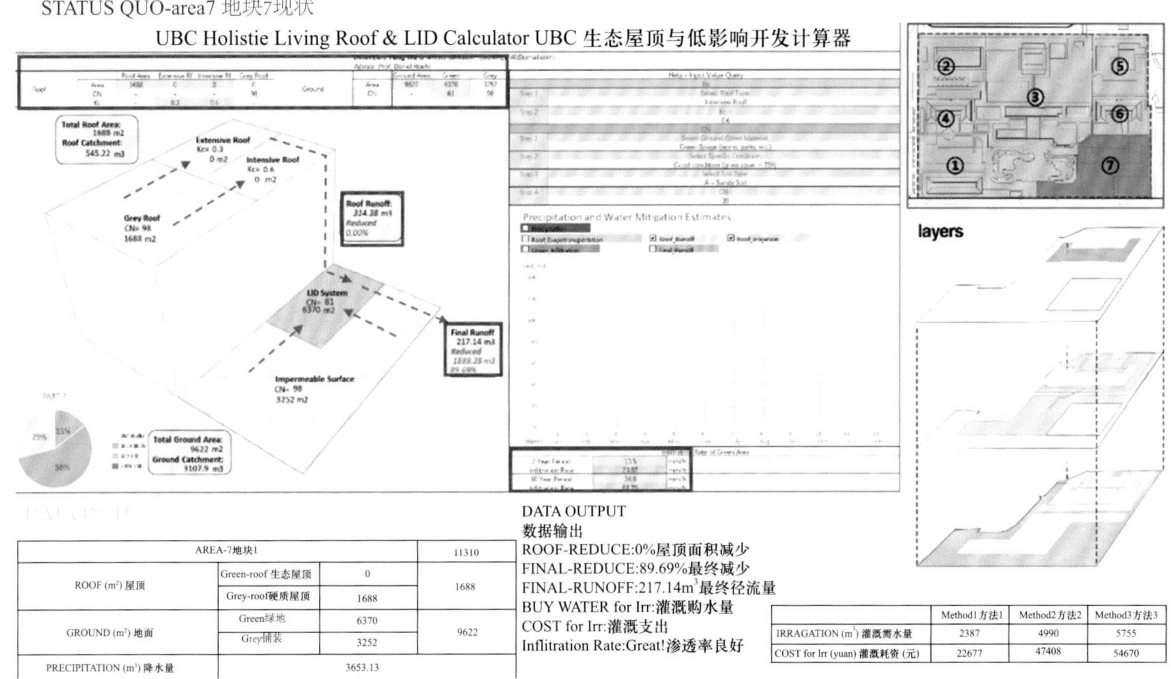

AREA-7地块1			11310
ROOF (m^2) 屋顶	Green-roof 生态屋顶	0	1688
	Grey-roof硬质屋顶	1688	
GROUND (m^2) 地面	Green绿地	6370	9622
	Grey铺装	3252	
PRECIPITATION (m^3) 降水量	3653.13		

	Method1方法1	Method2方法2	Method3方法3
IRRAGATION (m^3) 灌溉需水量	2387	4990	5755
COST for Irr (yuan) 灌溉耗资 (元)	22677	47408	54670

图 127　Site 7 Status assessment
地块七现状评估

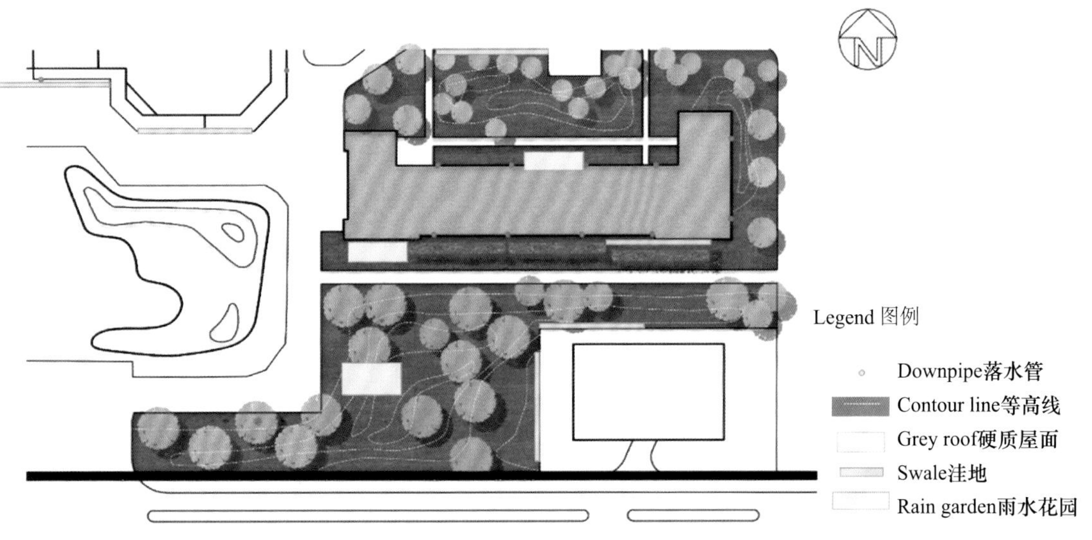

图 128　Site 7 Modified plan
地块七改进平面

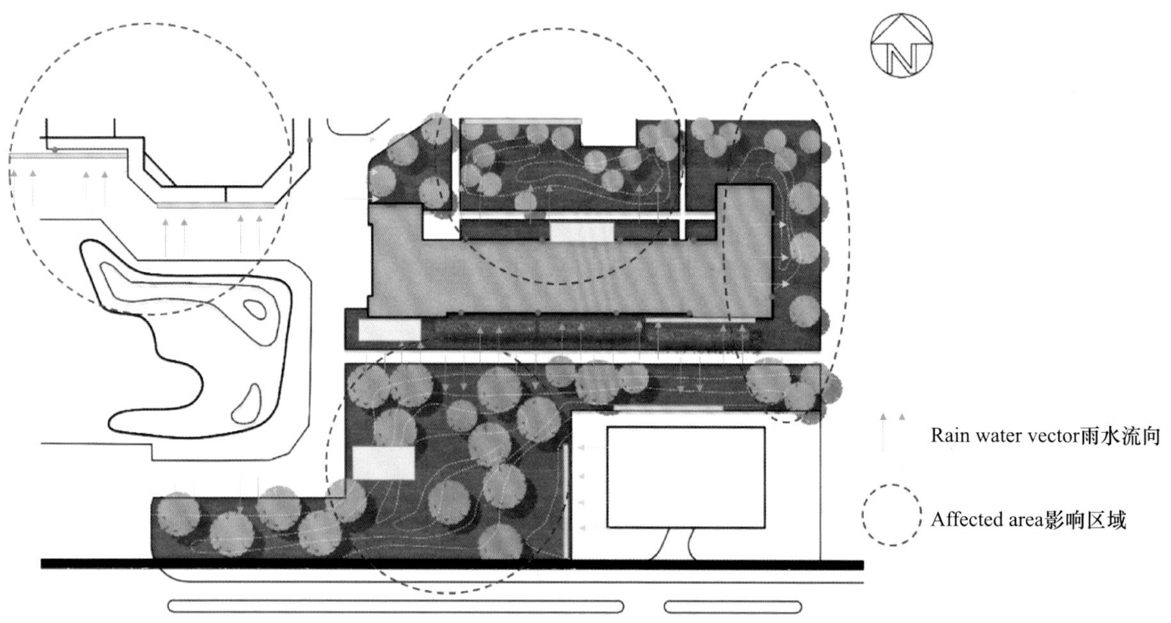

图 129　Site 4 Design analysis
地块七设计分析

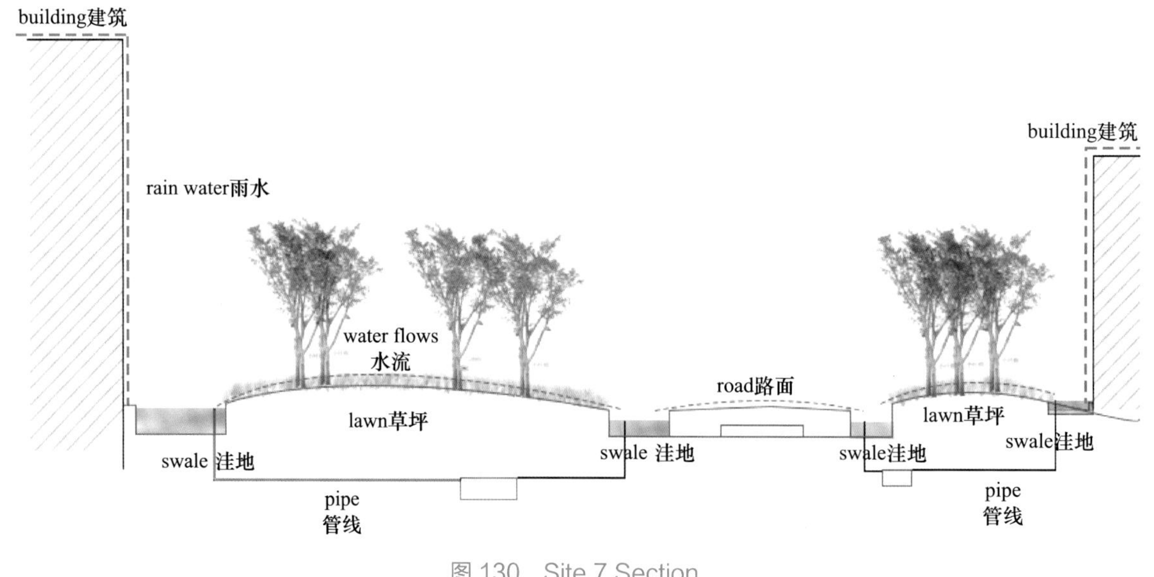

图 130　Site 7 Section
地块七剖面

3.2.9 Analysis of Design Result 设计结果分析

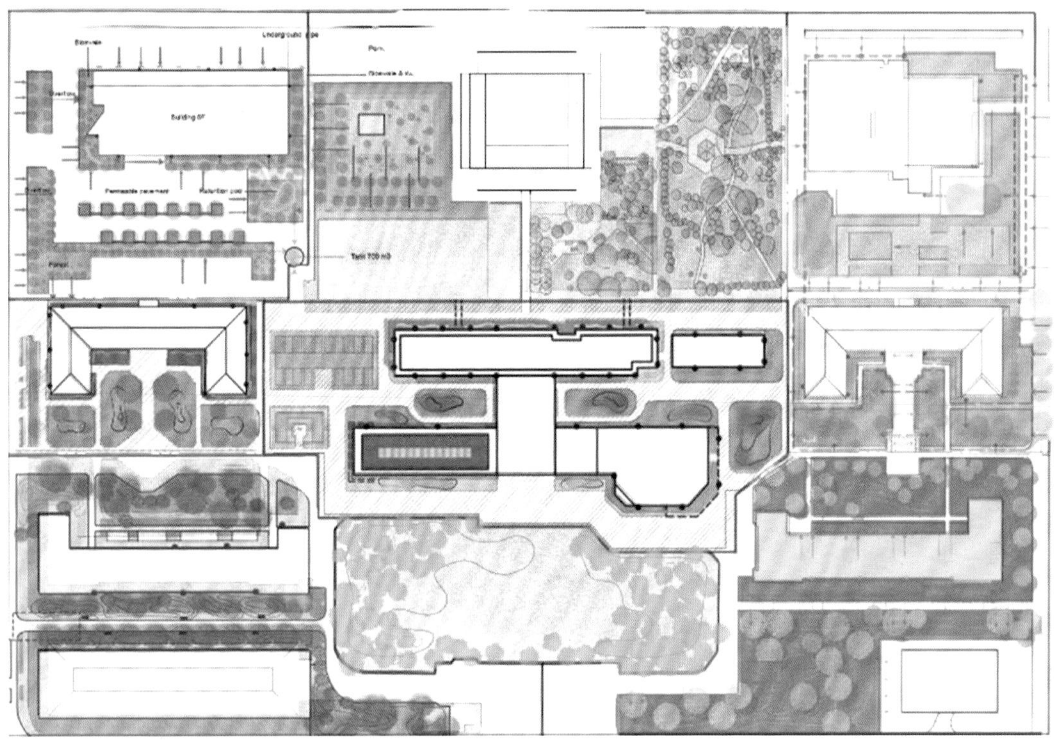

图 131 Master Plan
设计总平面

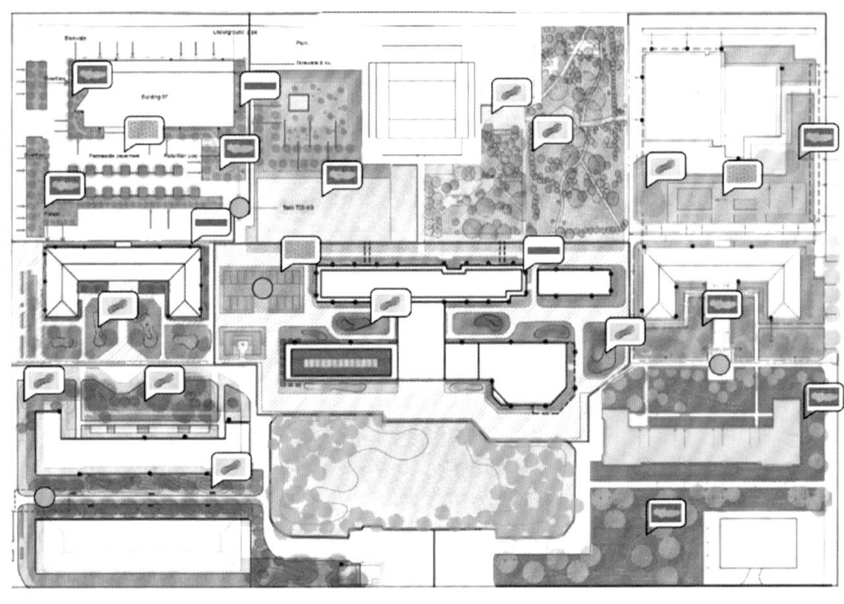

Legend 图例
- Downpipe落水管
- Open drain明沟
- Rain garden雨水花园
- Bioswale生态洼地
- Permeable paving透水铺装
- Impermeable paving不透水铺装
- Tank水箱

图 132 Master Plan with Details
细化总平面

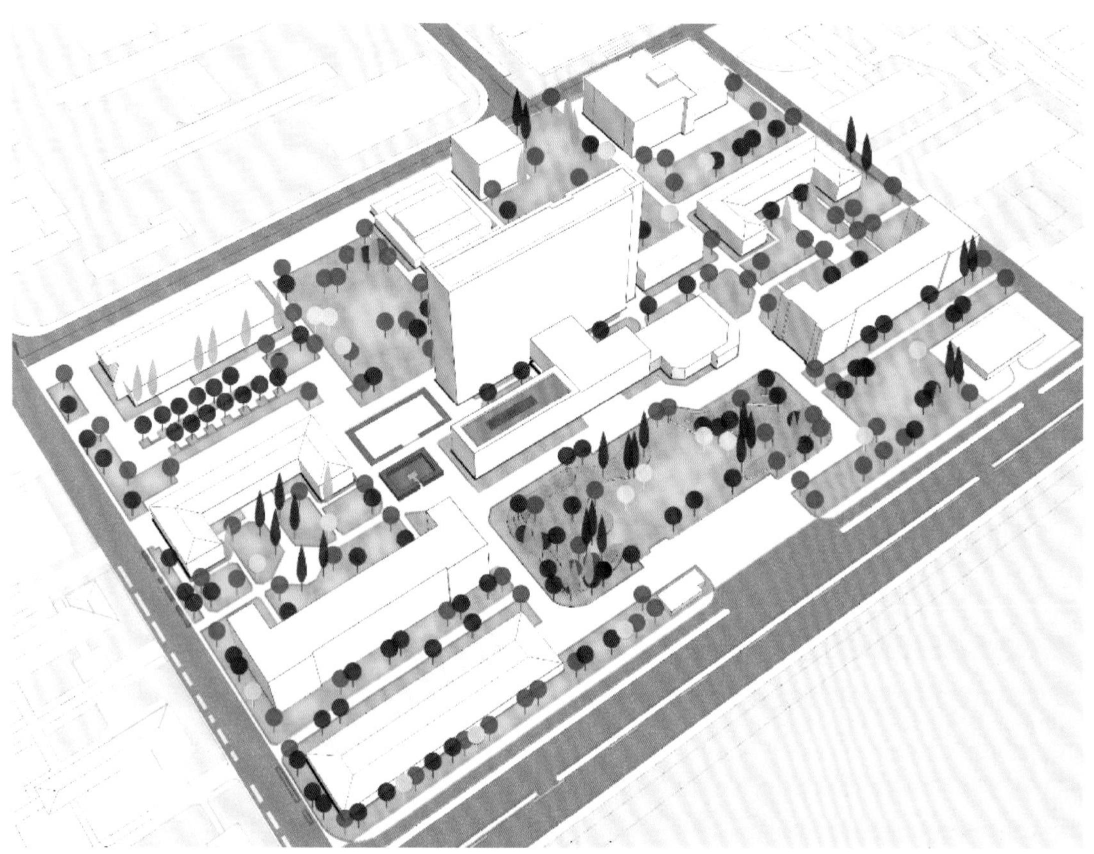

图 133 Aerial View
鸟瞰图

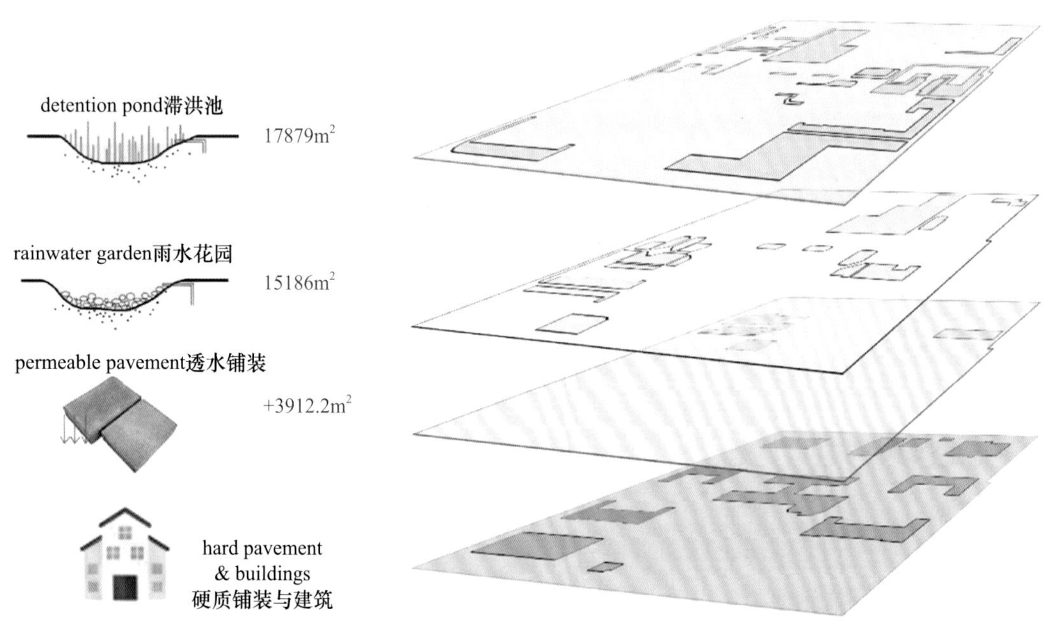

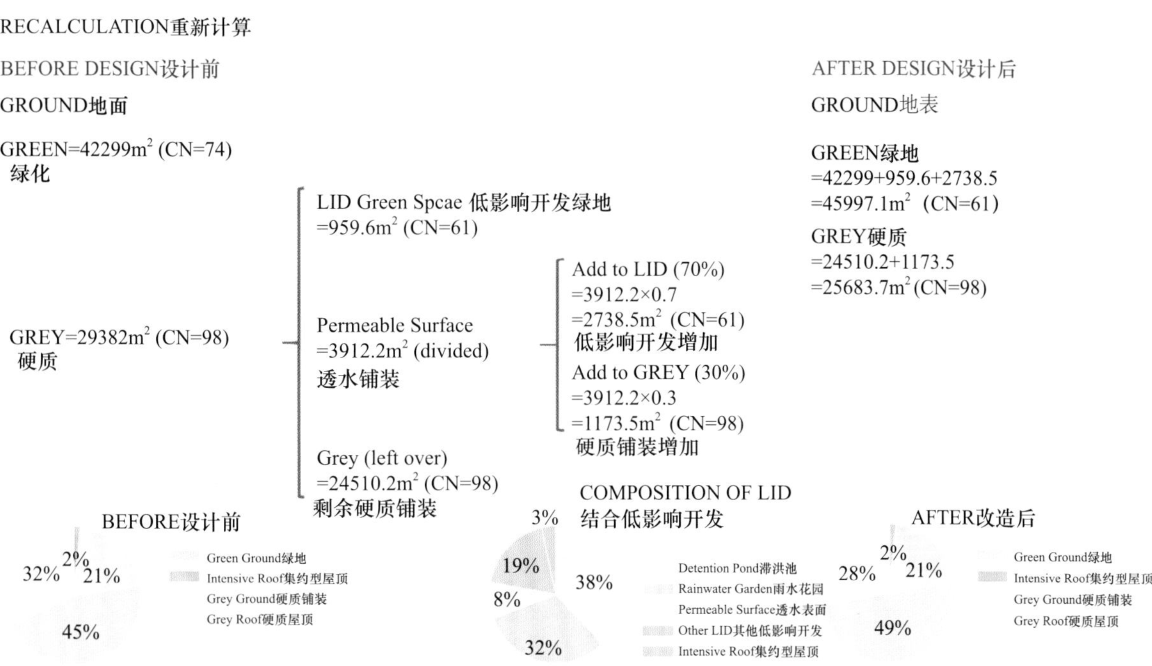

图 134 Analysis of Reconstruction
改造分析

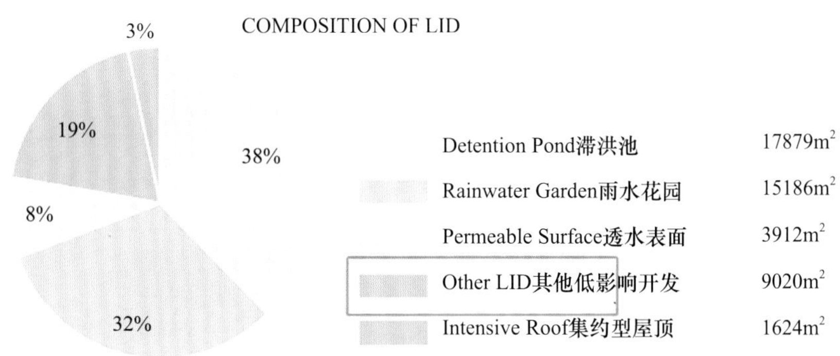

图 135 Composition of LID
低影响开发构成

NOTE:

We got the area by importing our master plan to AutoCAD and measuring the area roughly. As we had difficulty extracting the clear boundaries of each green space category, we left over an ambiguous part and named it ‘Other LID’. (Personally, I guess it contains a high percentage of bioswales and maybe some planting beds covered with gravels and wood stick…)

说明：

我们通过导入我们的总体规划获得了该区域到 AutoCAD 并粗略测量区域面积。因为难以提取每个类别绿色空间清晰的边界，我们便留下了一个模棱两可的部分并将其命名为“其他 LID”。（就个人而言，我们猜测它含有很高比例的生态洼地，也许还有用砾石和木条覆盖的一些种植床……）

Stay the same持平

AREA1地块1

CURRENT SITUATION 现状
ROOF-REDUCE: 0%屋顶减少
FINAL-REDUCE: 67.66%最终减少
FINAL-RUNOFF: 809.73m³最终径流量
BUY WATER for Irr: 6115m³灌溉额外需水量
COST for Irr: 5.8W RMB灌溉支出
Inflitration Rate: 77.87%下渗率

AFTER DESIGN设计后

ROOF-REDUCE: 0%屋顶减少
FINAL-REDUCE: 67.66%最终减少
FINAL-RUNOFF: 369.45m³最终径流量
BUY WATER for Irr: 5305.55m³灌溉额外需水量
COST for Irr: 5W RMB灌溉支出
Inflitration Rate: 77.87%下渗率

CONCLUSION结论

2500m³

FINAL RUNOFF REDUCE
440.28m³
0%
最终减少径流量

0.18×

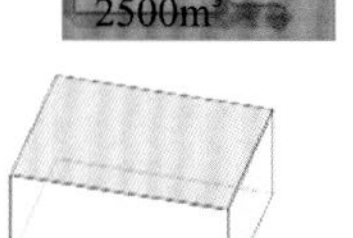

Olympic pool奥运泳池

BUY WATER FOR IRRIGATION REDUCE
810m³
0.8W RMB
灌溉额外需水量减少

2500m³

0.32×

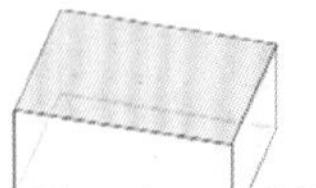

Olympic pool奥运泳池

INFLLTRATION RATE REDUCE 0%
下渗率降低 0%

AREA2地块2

+3100m² permeable pavement
增加3100m² 透水铺装

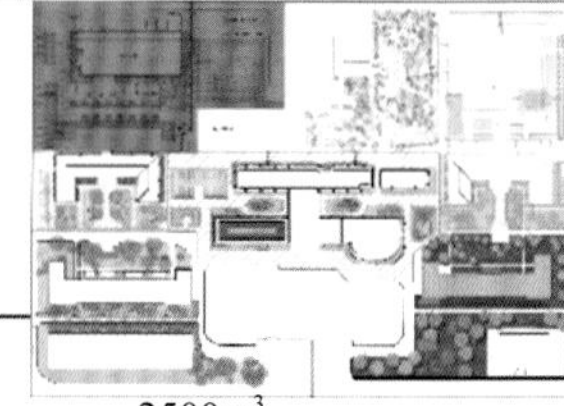

CURRENT SITUATION 现状
ROOF-REDUCE: 9.78%屋顶减少
FINAL-REDUCE: 53.99%最终减少
FINAL-RUNOFF: 1241.86m³最终径流量
BUY WATER for Irr: 3677.5m³灌溉额外需水量
COST for Irr: 3.49W RMB灌溉支出
Inflitration Rate: 129.15%下渗率

AFTER DESIGN设计后

ROOF-REDUCE: 9.78%屋顶减少
FINAL-REDUCE: 78.44%最终减少
FINAL-RUNOFF: 581.87m³最终径流量
BUY WATER for Irr: 3644.63m³灌溉额外需水量
COST for Irr: 3.46W RMB灌溉支出
Inflitration Rate: 104.31%下渗率

CONCLUSION结论

2500m³

FINAL RUNOFF REDUCE
581.87m³
78.44%
最终减少径流量

0.23×

Olympic pool奥运泳池

BUY WATER FOR IRRIGATION REDUCE
32.9m³
0.03W RMB
灌溉额外需水量减少

2500m³

0.013×

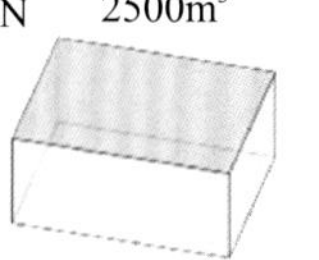

Olympic pool奥运泳池

INFILTRATION RATE REDUCE 24.84%
下渗率降低 24.84%

AREA3地块3

CURRENT SITUATION 现状
ROOF-REDUCE: 2.64%屋顶减少
FINAL-REDUCE: 67.55%最终减少
FINAL-RUNOFF: 1673.75m³最终径流量
BUY WATER for Irr: 12523.4m³灌溉额外需水量
COST for Irr: 11.9W RMB灌溉支出
Inflitration Rate: 78.42%下渗率

AFTER DESIGN设计后

ROOF-REDUCE: 2.64%屋顶减少
FINAL-REDUCE: 87.02%最终减少
FINAL-RUNOFF: 669.47m³最终径流量
BUY WATER for Irr: 11904.03m³灌溉额外需水量
COST for Irr: 11.3W RMB灌溉支出
Inflitration Rate: 71.5%下渗率

+787.02m² LID green ground
+574.6m² permeable pavement
增加787.02m²低影响开发绿地
增加574.6m²透水铺装

CONCLUSION结论

FINAL RUNOFF REDUCE
1004.28m³
19.47%
最终减少径流量

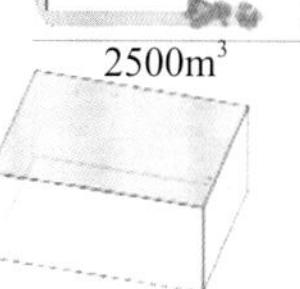

2500m³
0.4×
Olympic pool奥运泳池

BUY WATER FOR IRRIGATION REDUCE
619.37m³
0.6W RMB
灌溉额外需水量减少

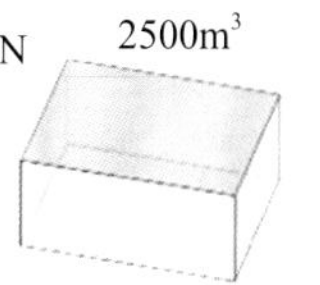

2500m³
0.25×
Olympic pool奥运泳池

INFILTRATION RATE REDUCE 6.92%
下渗率降低 6.92%

AREA4地块4

CURRENT SITUATION 现状
ROOF-REDUCE: 0%屋顶减少
FINAL-REDUCE: 66.02%最终减少
FINAL-RUNOFF: 318.08m³最终径流量
BUY WATER for Irr: 1692.4m³灌溉额外需水量
COST for Irr: 1.6W RMB灌溉支出
Inflitration Rate: 98.37%下渗率

AFTER DESIGN设计后

ROOF-REDUCE: 0%屋顶减少
FINAL-REDUCE: 83.94%最终减少
FINAL-RUNOFF: 150.31m³最终径流量
BUY WATER for Irr: 1616.19m³灌溉额外需水量
COST for Irr: 1.5W RMB灌溉支出
Inflitration Rate: 8265%下渗率

+172.39m² LID green ground
+237.6m² permeable pavement
增加172.39m²低影响开发绿地
增加237.6m²透水铺装

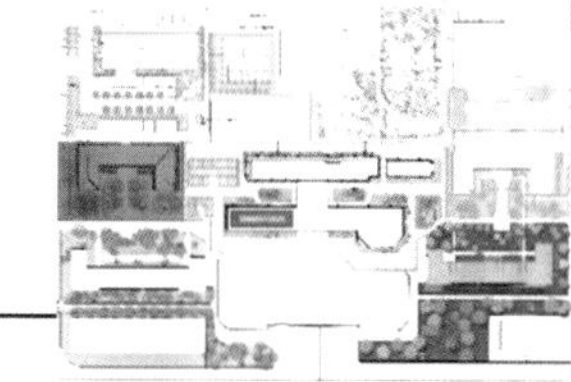

CONCLUSION结论

FINAL RUNOFF REDUCE
167.77m³
17.92%
最终减少径流量

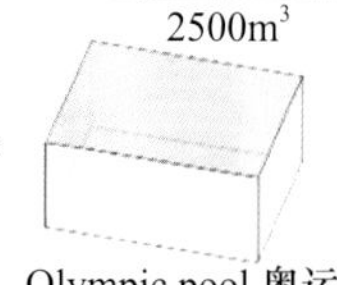

2500m³
0.07×
Olympic pool 奥运泳池

BUY WATER FOR IRRIGATION REDUCE
76.21m³
724 RMB
灌溉额外需水量减少

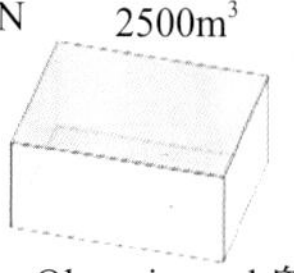

2500m³
0.03×
Olympic pool 奥运泳池

INFILTRATION RATE REDUCE 15.72%
下渗率降低 15.72%

AREA5地块5

CURRENT SITUATION 现状
ROOF-REDUCE: 0%屋顶减少
FINAL-REDUCE: 60.8%最终减少
FINAL-RUNOFF: 633.76m^{3}最终径流量
BUY WATER for Irr: 2971.8m^{3}灌溉额外需水量
COST for Irr: 2.82W RMB灌溉支出
Inflitration Rate: 9.49%下渗率

AFTER DESIGN设计后

ROOF-REDUCE: 0%屋顶减少
FINAL-REDUCE: 79.44%最终减少
FINAL-RUNOFF: 332.35m^{3}最终径流量
BUY WATER for Irr: 1419.66m^{3}灌溉额外需水量
COST for Irr: 1.35W RMB灌溉支出
Inflitration Rate: 9.49%下渗率

CONCLUSION结论

FINAL RUNOFF REDUCE
332.35m^3
79.44%
最终减少径流量

0.13×
2500m^3
Olympic pool 奥运泳池

BUY WATER FOR IRRIGATION REDUCE
1552m^3
1.47W RMB
灌溉额外需水量减少

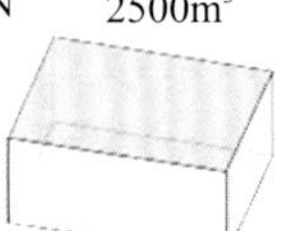
0.62×
2500m^3
Olympic pool 奥运泳池

INFILTRATION RATE REDUCE 0%
下渗率降低 0%

Stay the same保持不变

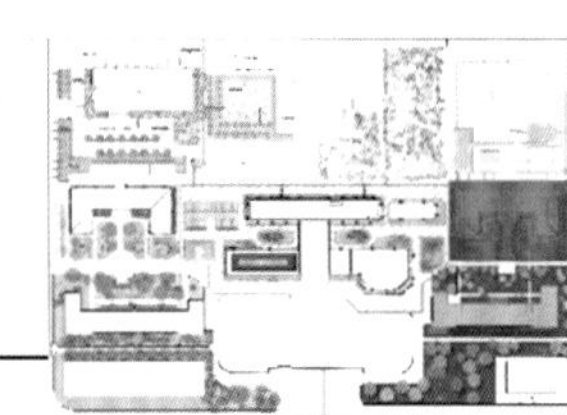

AREA6地块6

CURRENT SITUATION 现状
ROOF-REDUCE: 0%屋顶减少
FINAL-REDUCE: 6.92%最终减少
FINAL-RUNOFF: 331.77m^{3}最终径流量
BUY WATER for Irr: 2374.9m^{3}灌溉额外需水量
COST for Irr: 2.26W RMB灌溉支出
Inflitration Rate: 79.9%下渗率

AFTER DESIGN设计后

ROOF-REDUCE: 0%屋顶减少
FINAL-REDUCE: 84.66%最终减少
FINAL-RUNOFF: 153.8m^{3}最终径流量
BUY WATER for Irr: 2041.2m^{3}灌溉额外需水量
COST for Irr: 1.94W RMB灌溉支出
Inflitration Rate: 79.9%下渗率

CONCLUSION结论

FINAL RUNOFF REDUCE
153.8m^3
84.66%
最终减少径流量

0.06×
2500m^3
Olympic pool 奥运泳池

BUY WATER FOR IRRIGATION REDUCE
333.7m^3
0.32W RMB
灌溉额外需水量减少

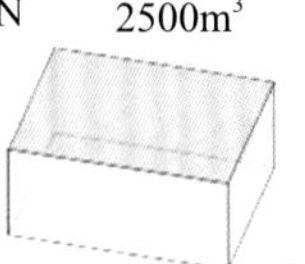
0.13×
2500m^3
Olympic pool 奥运泳池

INFILTRATION RATE REDUCE 0%
下渗率降低 0%

AREA7地块7

CURRENT SITUATION 现状
ROOF-REDUCE: 0%屋顶减少
FINAL-REDUCE: 73.96%最终减少
FINAL-RUNOFF: 548.45m³最终径流量
BUY WATER for Irr: 6723m³灌溉额外需水量
COST for Irr: 6.39W RMB灌溉支出
Inflitration Rate: 61.79%下渗率

AFTER DESIGN设计后

ROOF-REDUCE: 0%屋顶减少
FINAL-REDUCE: 89.69%最终减少
FINAL-RUNOFF: 217.14m³最终径流量
BUY WATER for Irr: 5640m³灌溉额外需水量
COST for Irr: 5.36W RMB灌溉支出
Inflitration Rate: 61.79%下渗率

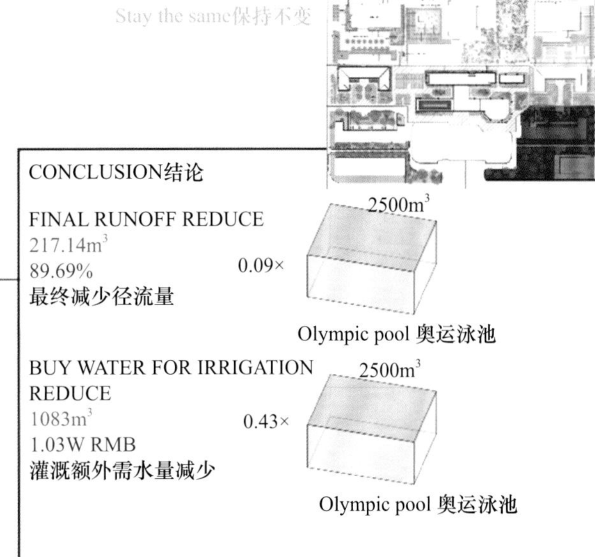

CONCLUSION结论

FINAL RUNOFF REDUCE
217.14m³
89.69%
最终减少径流量

BUY WATER FOR IRRIGATION REDUCE
1083m³
1.03W RMB
灌溉额外需水量减少

INFILTRATION RATE REDUCE 0%
下渗率降低 0%

AREA-ALL整个场地

UBC Holistie Living Roof & LID Calculator UBC生态屋顶与低影响开发计算器

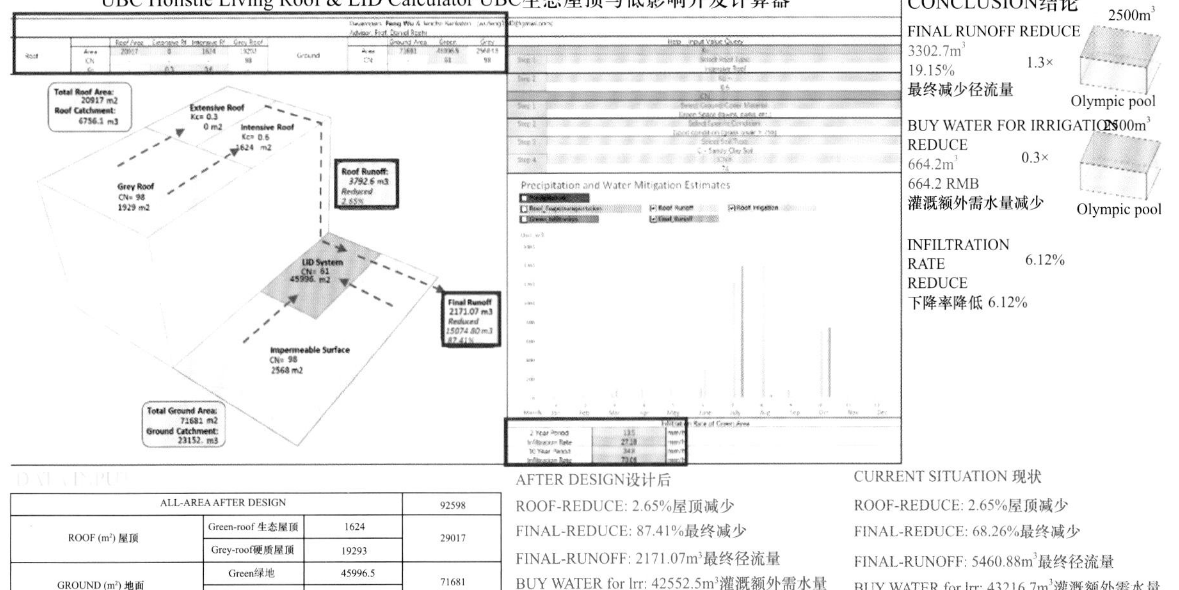

CONCLUSION结论

FINAL RUNOFF REDUCE
3302.7m³
19.15%
最终减少径流量
2500m³
1.3×
Olympic pool

BUY WATER FOR IRRIGATION REDUCE
664.2m³
664.2 RMB
灌溉额外需水量减少
2500m³
0.3×
Olympic pool

INFILTRATION RATE REDUCE 6.12%
下降率降低 6.12%

ALL-AREA AFTER DESIGN			92598
ROOF (m²) 屋顶	Green-roof 生态屋顶	1624	29017
	Grey-roof硬质屋顶	19293	
GROUND (m²) 地面	Green绿地	45996.5	71681
	Grey铺装	25584.5	
PRECIPITATION (m³) 降水量	29909.15		

AFTER DESIGN设计后
ROOF-REDUCE: 2.65%屋顶减少
FINAL-REDUCE: 87.41%最终减少
FINAL-RUNOFF: 2171.07m³最终径流量
BUY WATER for Irr: 42552.5m³灌溉额外需水量
COST for Irr: 40.4W RMB灌溉支出
Inflitration Rate: 70.06%下渗率

CURRENT SITUATION 现状
ROOF-REDUCE: 2.65%屋顶减少
FINAL-REDUCE: 68.26%最终减少
FINAL-RUNOFF: 5460.88m³最终径流量
BUY WATER for Irr: 43216.7m³灌溉额外需水量
COST for Irr: 41.1W RMB灌溉支出
Inflitration Rate: 76.18%下渗率

3.2.10 Conclusions 结论

1. WHAT WE GOT? 我们得到了什么？

图 136　Flood in campus
校园洪水事件

Currently we have 42299m² green space (contains several rainwater gardens) and 1624m² intensive roof. But they are not functioning efficiently for lacking maintainance.

We unavoidably have flooding risk when there's a short-duration high-intensity rainfall on campus.

That is a thorny problem because future climate change could trigger more frequent and extreme weather events that we are not equipped to manage at present.

2. WHAT WE CAN GET? AND HOW MUCH IT WILL COST?

目前我们拥有 42299 平方米的绿地（包含几个雨水花园）和 1624 平方米的密集屋顶。但由于缺乏维护，它们无法有效运作。

当校园内出现短时高强度降雨时，我们不可避免地会有洪水风险。

这是一个棘手的问题，因为未来的气候变化可能引发我们目前无法掌控的更频繁和极端的天气事件。

2. 我们可以得到什么？需要多少钱？

图 137　Rendering of campus garden after transformation
改造后校园花园效果图

AN ATTRACTIVE AND FUNCTIONAL SPONGE CAMPUS
一个有吸引力的、功能丰富的海绵校园

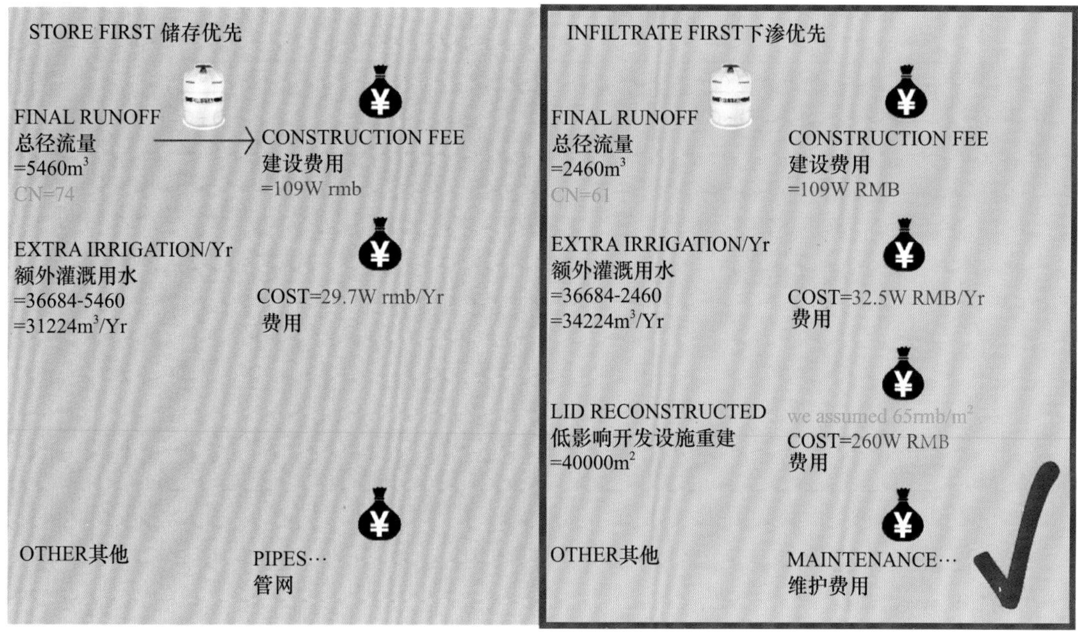

CONCLUSION 结论

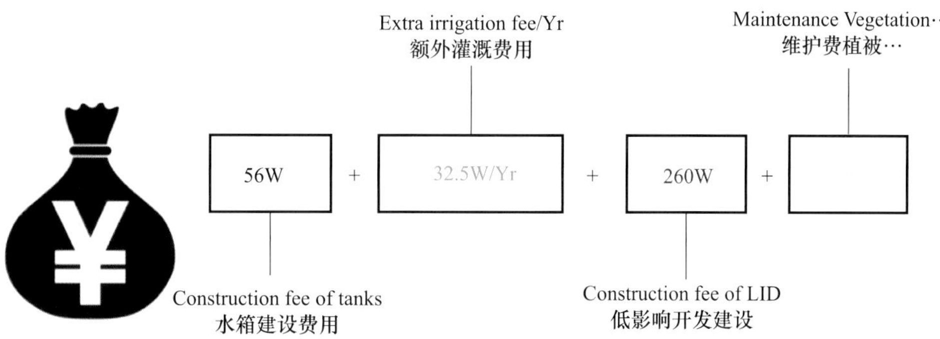

COSTLY INDEED…
确实价格不菲

IT'S A COSTLY RROJECT INDEED.
它的确是一个高花费的项目

AND WITH ALL THIS MONEY,
用这些钱

WHAT DO WE GET IN RETURN?
我们能获得什么回报?

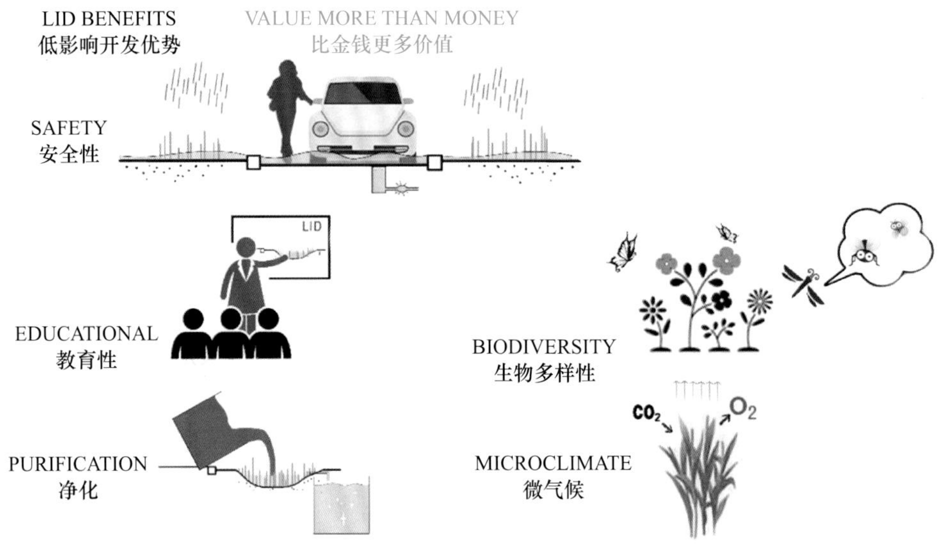

WHAT SHOULD WE DO IN THE FUTURE？我们未来要做什么？

MAINTAINANCE维护

we need我们需要

professional horticulturist专业的园艺学家

technical worker技工

money钱

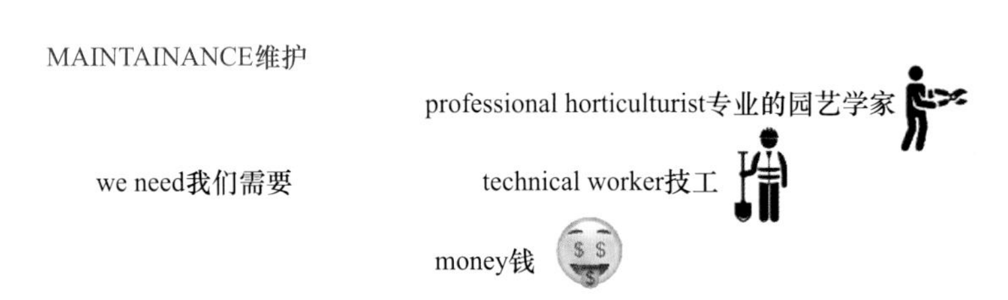

EDUCATE THE PUBLIC公众教育

we can use我们可以利用

short video短片

lecture讲座

public service advertisement公益广告

HOW TO MAINTAIN IN THE FUTURE？未来如何维护？

WEEKLY INSPECTING每周检查

problem: not infiltrating within 24h
问题：24小时内未下渗

problem: gullying after rain event
问题：雨后形成冲沟

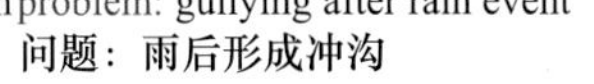

problem: invasive plants
问题：植物入侵

note: observe the rain garden during rainfall and note any success
注意：下雨时观察雨水花园并记录是否成功

MULCHING

note: add mulch every spring to maintain a 3 inch mulch layer
注意：每年春天补充护根覆盖物以保持3英寸的覆盖物层

PRUNING

note: cut back dead vegetation，flowers and tattered plants
注意：修剪枯萎的植物、花朵和被毁坏的植物

HOW TO EDUCATE THE PUBLIC? 如何进行公众教育?

SHORT VIDEO短片

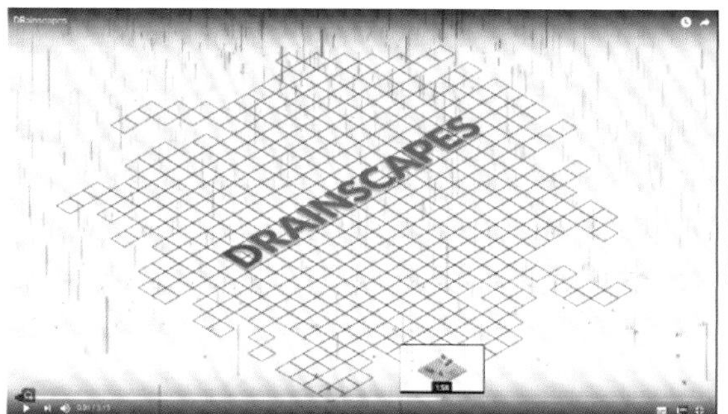

PUBLIC SERVICE ADVERTISEMENT公益广告

NOTE:
About the public service advertisement's part, we failed to find a vivid ad related to LID. so we simply use this one instead, and maybc that's a good opportunity for us to design our own ads.

注:
关于公益广告部分我们没有找到一个LID相关的生动广告，所以我们只是使用了这个广告，也许这是我们设计自己的广告的好机会。

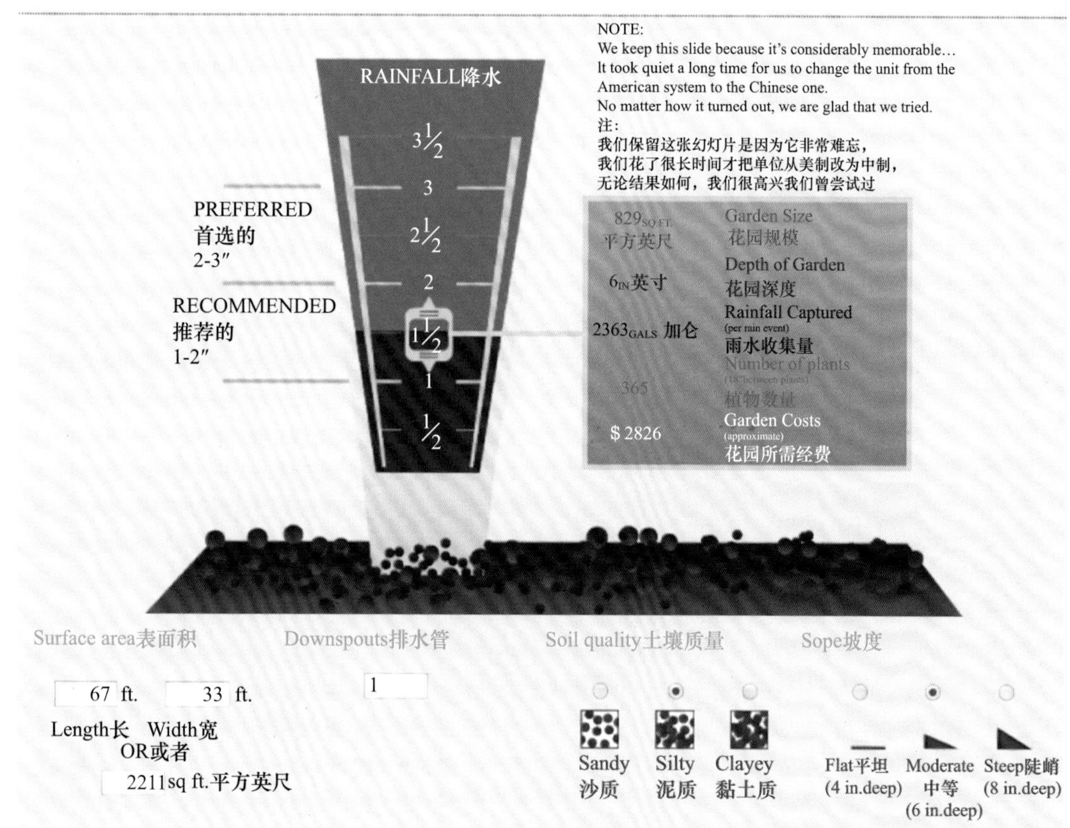

LID in a New Ecological Community
高碑店某居住区景观雨洪管理综合改造

Project date:August 2018 项目日期：2018 年 8 月

1. Site condition analysis 场地条件分析

1.1 Site basic information 场地基本信息

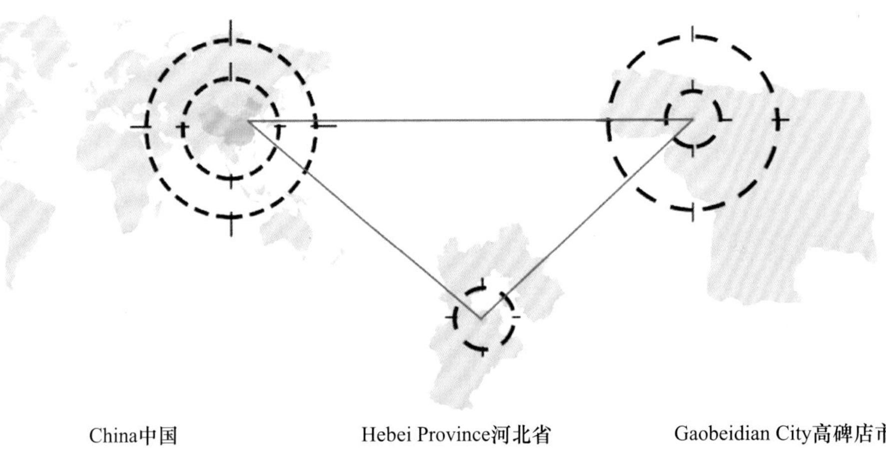

图 1 Location of site
场地区位

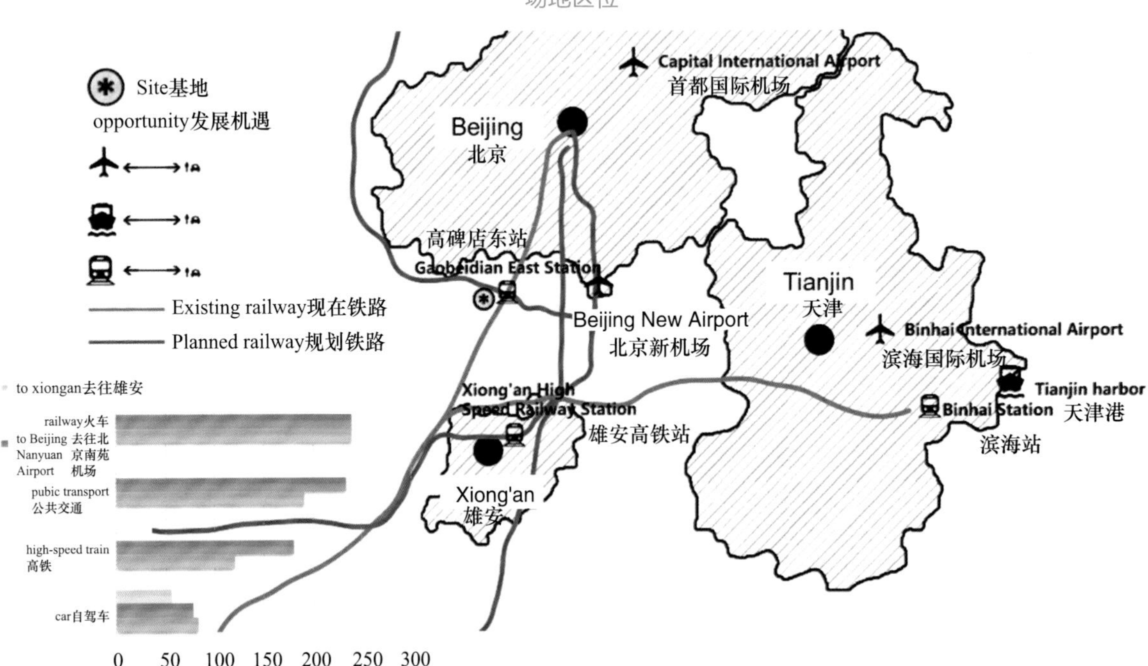

图 2 Relation with surrounding cities
场地与周边城市的联系

图 3　Site cultural origin
场地文化起源

Designers sought the natural and cultural heritage of the region to handle the relationship with history.
设计师通过搜寻该地区的自然和文化遗产来处理场地与历史的关系。
Ziquan Palace is the living place where the emperors of the Qing Dynasty visited the south.
紫泉宫是清朝皇帝南巡的住所。

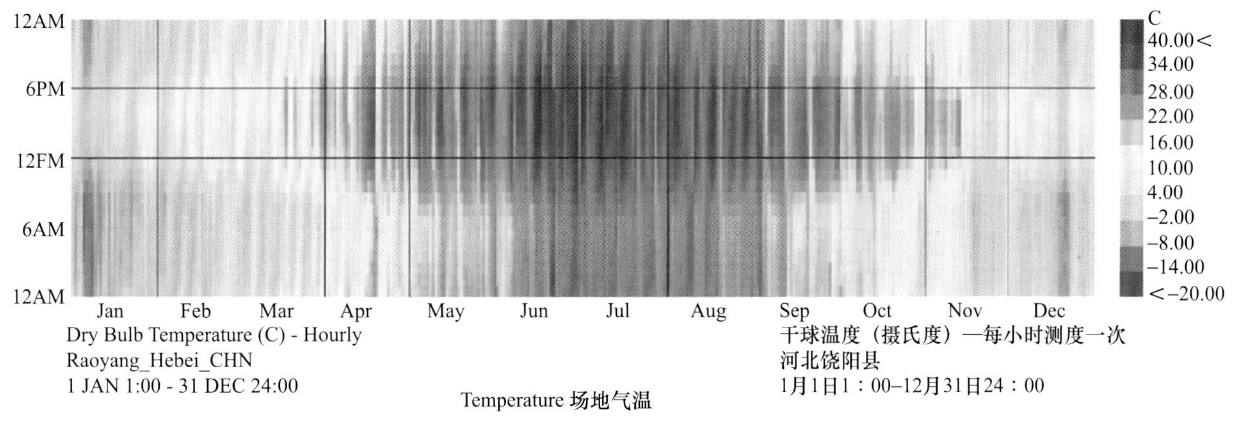

图 4　Temperature
场地气温

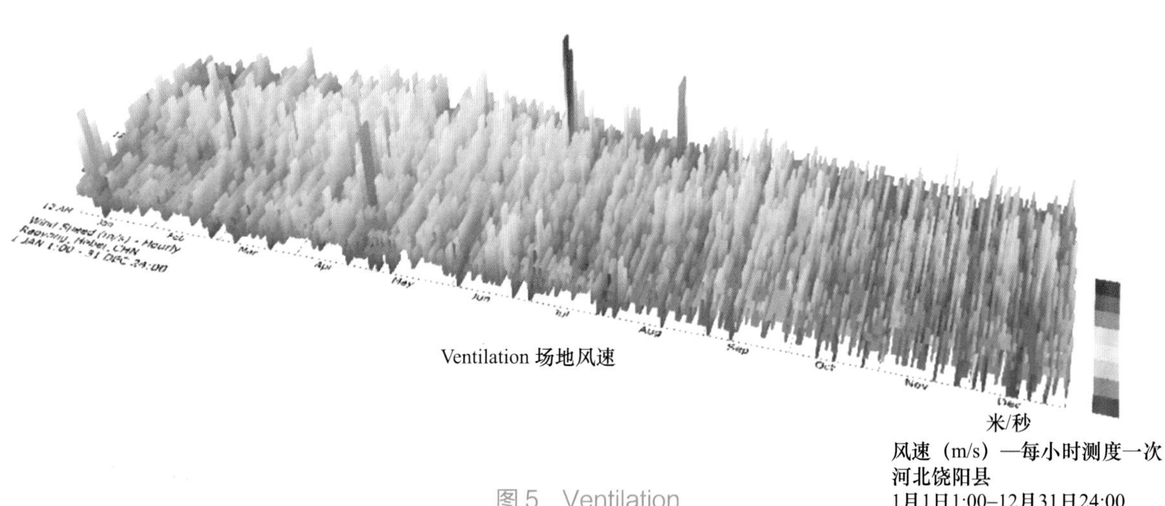

图 5　Ventilation
场地风速

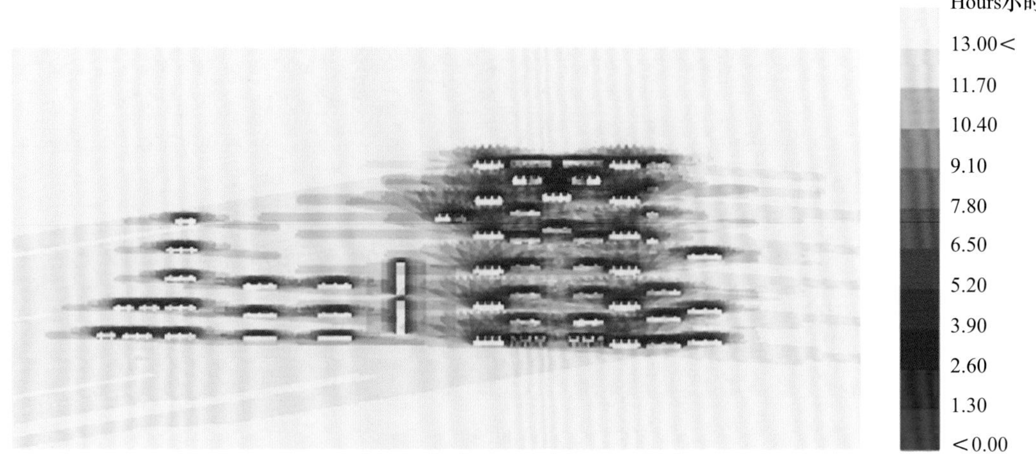

图 6 Sunlight hours
场地日照

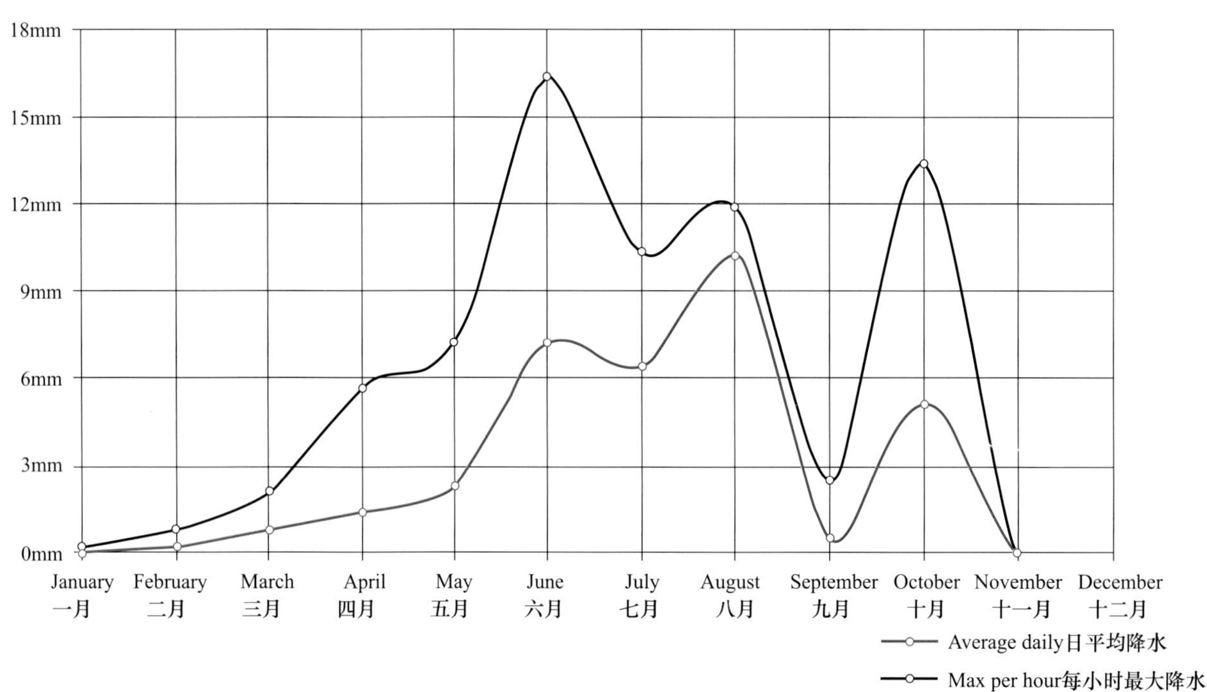

图 7 Average precipitation
场地平均降水

Gaobeidian City is located in the central part of Hebei Province, southwest of Beijing. It is located in the hinterland of Beijing, Tianjin and Baoding, and is surrounded by the capital Beijing-Tianjin Economic Circle. It is an important industrial city to north of Baoding and south of Beijing.

The site belongs to temperate monsoon climate. The area's annual average temperature is 12℃ , the highest is 43℃ in July ,the lowest is -18℃ in Janaury

高碑店市位于河北省中部，北京西南部，地处北京、天津、保定三角腹地，环首都京津经济圈，是京南保北重要的工业城市。

场地属于温带季风气候，年平均气温 12℃，年最高气温在 7 月为 43℃，年最低气温为 1 月的—18℃。由于周边居住建筑的遮挡，场地通风

.Due to the obstruction of the surrounding residential buildings, the ventilation of the site is not good, but the sunshine is sufficient.

状况不佳，但日照充足。降水量不大，主要集中于夏季。

1.2 Site current situation analysis 场地现状分析

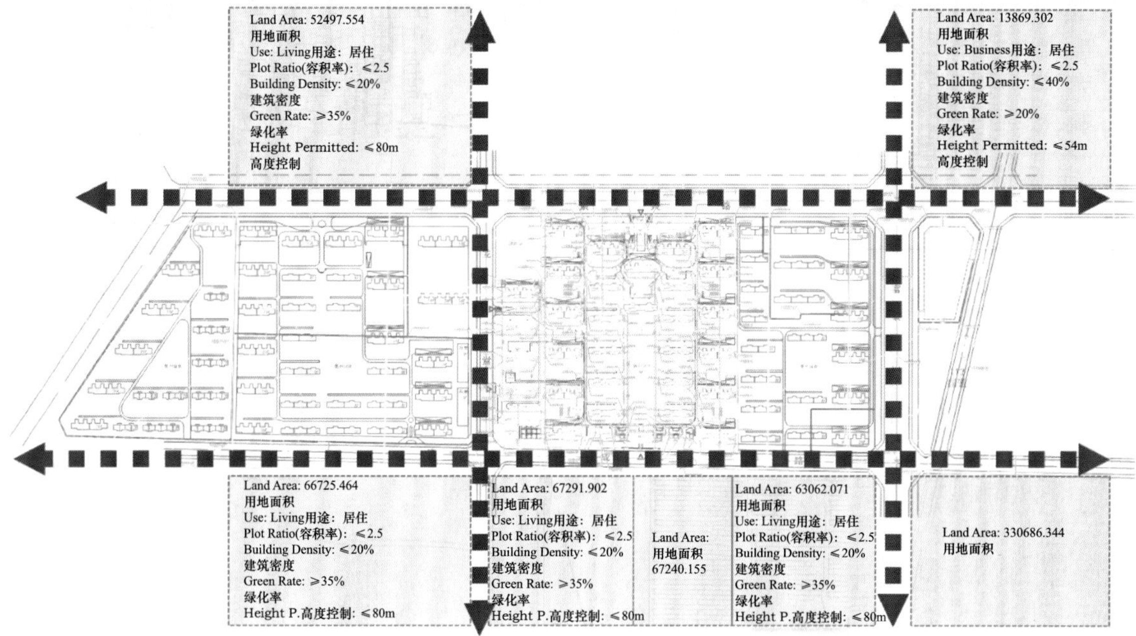

图 8 Surrounding land distribution
周边用地布局

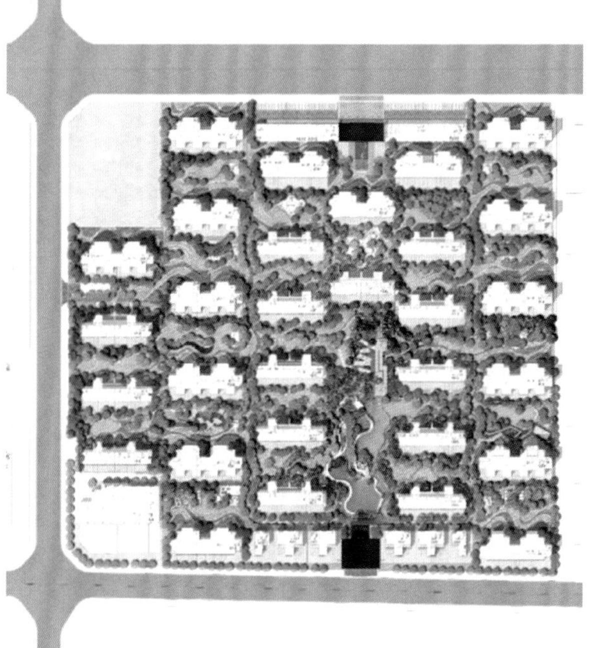

NAME: Gaobeidian LID Project

POPULATION: About 11,000 People

Here is the design plan of the project,which has obvious transportation system, centural park and several green sites between the buildings. The beautiful part should be retained and we try to add LID measures to control the stormwater.

项目名称：高碑店生态改造项目

人口：约 1.1 万

这是该项目的设计方案，具有完善的交通系统、中心公园和建筑物之间的几处绿地空间。我们的设计将应保留原方案的优秀部分，并尝试增加 LID 措施来控制雨水。

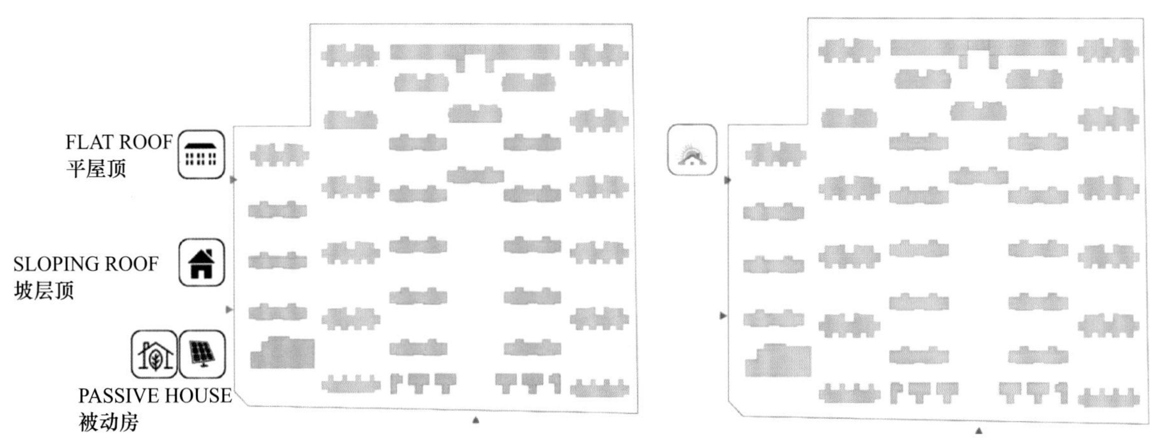

图 9　Building category
建筑类别

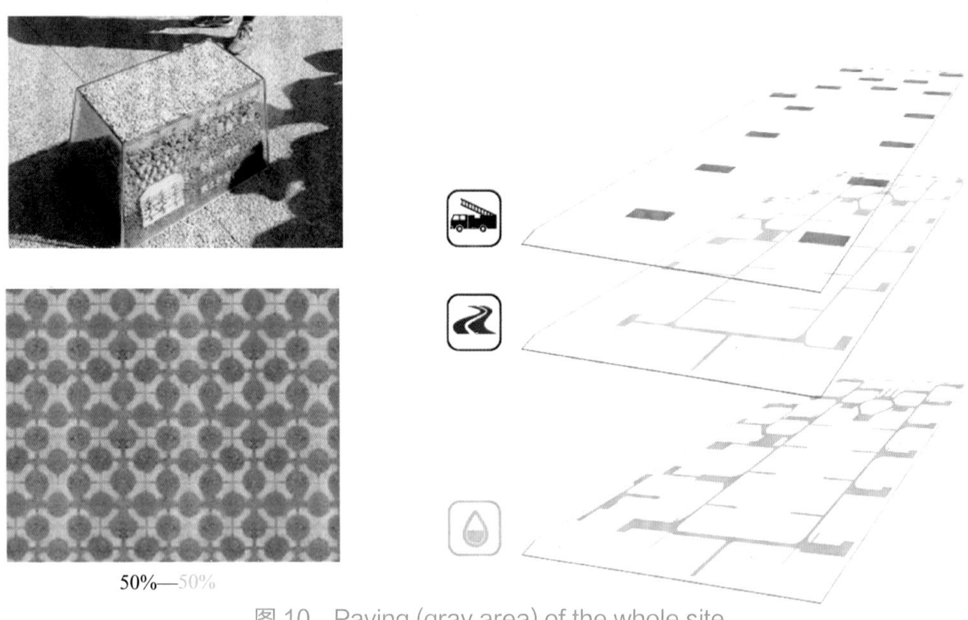

图 10　Paving (gray area) of the whole site
场地铺装（灰色区域）

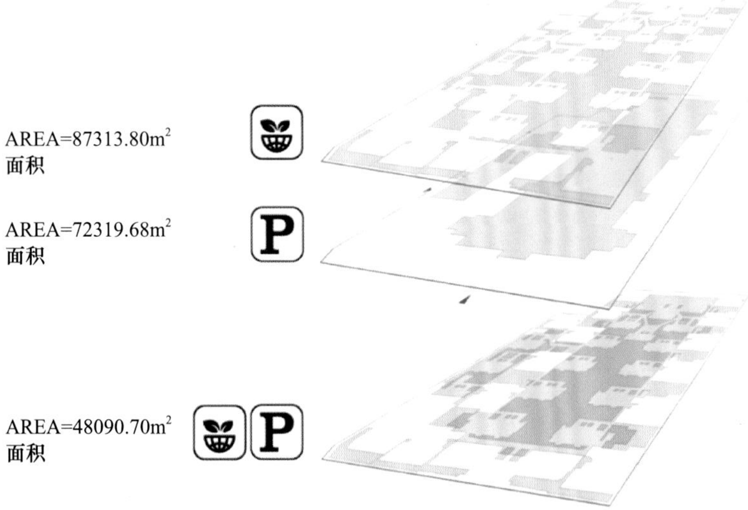

图 11　Green space analysis
绿地分析

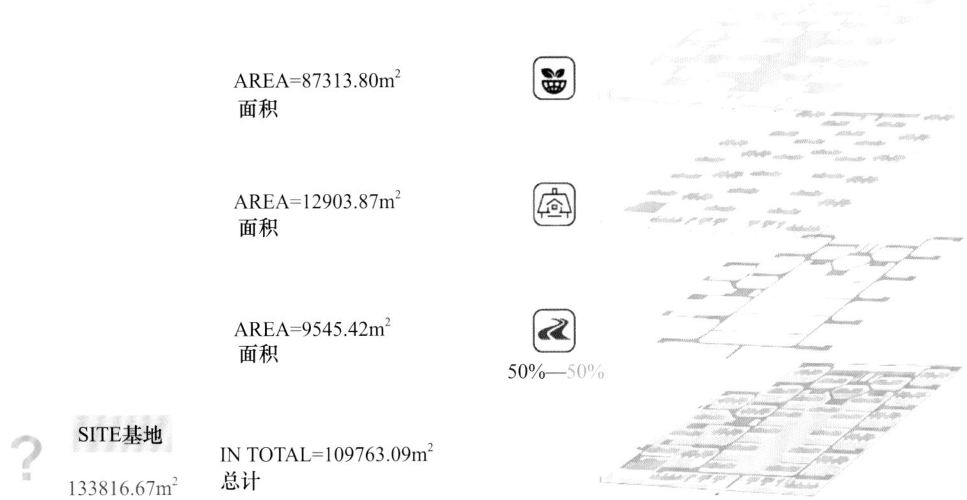

图 12　Available green space in total
可用的绿色空间

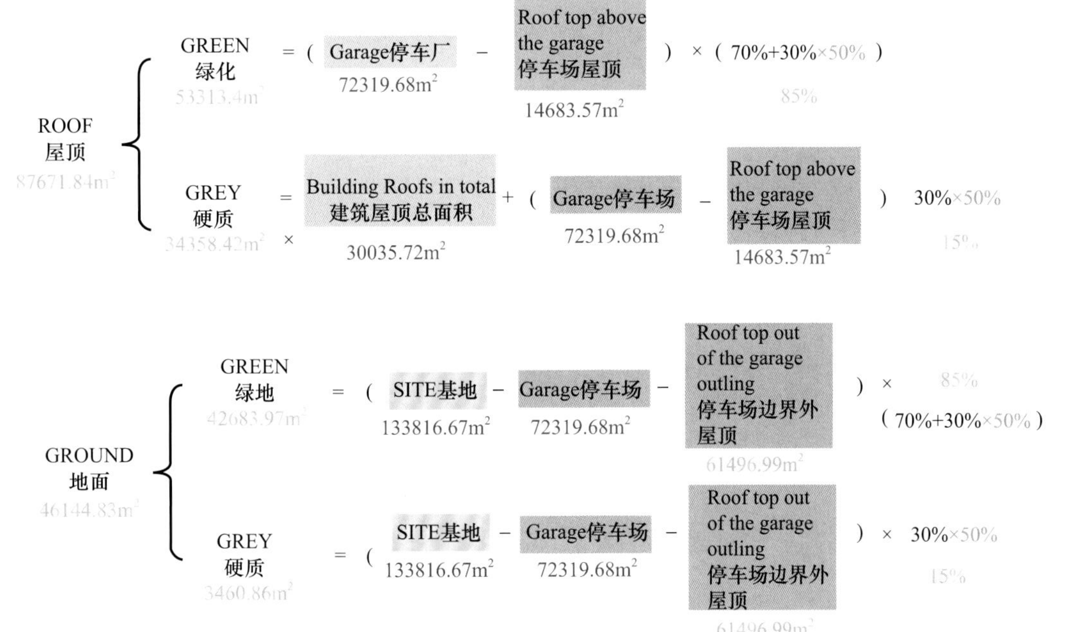

图 13　Logic of calaclation
计算方法

1.3 Calculation based on existing plan 根据已有平面图计算

1.3.1 Chinese calculation method 中式方法

表 1 statistics of surface types and area of the Site
场地表面类型与面积统计

The type of surface 表面类型	Area (m^2) 面积
Green Roof 生态屋顶	0
Pavement 铺装	64163.57
Green 绿化	17396.65
Hard Roof 硬质屋面	26071.3
Basement Green 下沉绿地	26900.34
Water 水面	900

表 2 Building economic and technical indicators
建筑经济技术指标

Indicator 指标	Rate (%) 数值
Green 绿化	32.90%
Building Area 建筑	19.38%
Hard Pavement 硬质铺装	47.72%

Green=Green+Basement Green+Water 绿地 = 绿地 + 下沉绿地 + 水面
Building Area=Hard Roof+Green Roof 建筑 = 硬质屋顶 + 生态屋顶
Hard Pavement=Pavement 硬质铺装 = 铺装

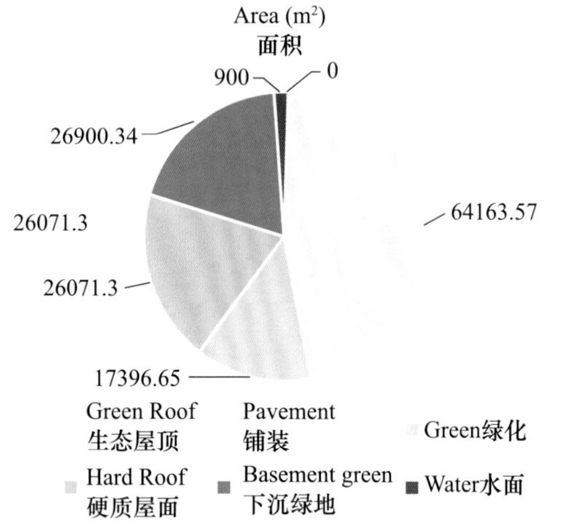

图 14 Pie chart of surface types and area of the site
场地表面类型与面积饼状图

Rate (%)
数值
32.90%
47.72%
19.38%
Green绿化
Building Area
绿化建筑
Hard Pavement
硬质铺装

图 15 Pie chart of building economic and technical indicators
建筑经济技术指标饼状图

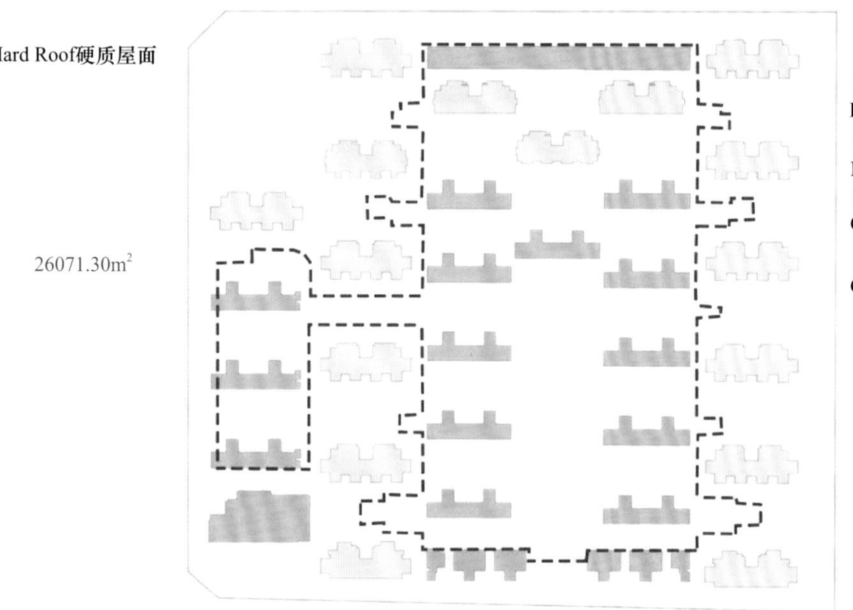

图 16 Hard roof
硬质屋面

Green, Water, Pavement 绿地、水系、铺装

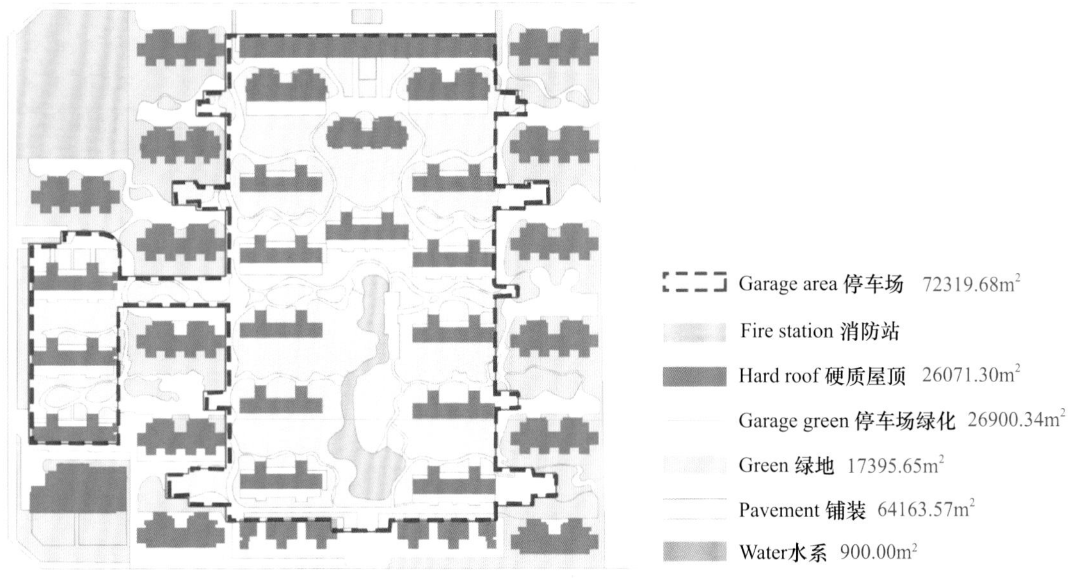

图 17 Green space, water, pavement analysis
绿地、水系、铺装分析

表 3 Calculation of final runoff of the site in 10-year periodand 3-year period
场地 10 年期与 3 年期总径流量计算

current 现状 type 类型	area (m^2) 面积	runoff coefficient 径流系数		W (runoff/m^3) 总径流量 10-year period（10 年期）	 3-year period（3 年期）
green roof 生态屋顶	0.00	0.30		0.00	0.00
road (impervious) 不透水路面	64163.57	0.90		12069.17	6236.70
green land 绿地	17396.65	0.15		545.38	281.83
grey roof 硬质屋顶	9502.60	0.90		1787.44	923.65
basement which is covered with green land (green land above the garage) 绿化覆盖的地下室（停车场屋顶绿化）	26900.34	0.15		843.33	435.79
			sum	15245.32	7877.97

Final ruoff 总径流量 **15245.32m³** **7877.97m³**

$$W=10\psi_{ZC}h_yF$$

W——The total runoff (m^2) 总径流量；

ψ_{ZC}——Rainfall comprehensive runoff coefficient 雨水综合径流系数；

h_y——Designed rainfall (mm) 设计降水量；

F——Catchment area (hm^2) 集水面积。

1.3.2 UBC Calculation method UBC 计算方法

表 4 Statistics of surface types and area in the site
场地表面类型与面积统计

The type of surface 表面类型		Area (m^2) 面积
Roof 屋顶	Green Roof 绿色屋顶	26900.34
	Grey Roof 硬质屋顶	59291.54
Ground 地面	Green 绿化	17395.65
	Grey 铺装	30943.33

表 5 Area ratio of roof to ground in the site
场地屋顶与地面面积比例

The type of surface 表面类型	Rate (%) 比例
Roof 屋顶	64.06%
Ground 地面	35.94%

Roof=Green+Roof+Grey Roof 屋顶 = 绿化 + 屋顶 + 硬质屋顶
Ground=Green+Grey 地面 = 绿化 + 铺装

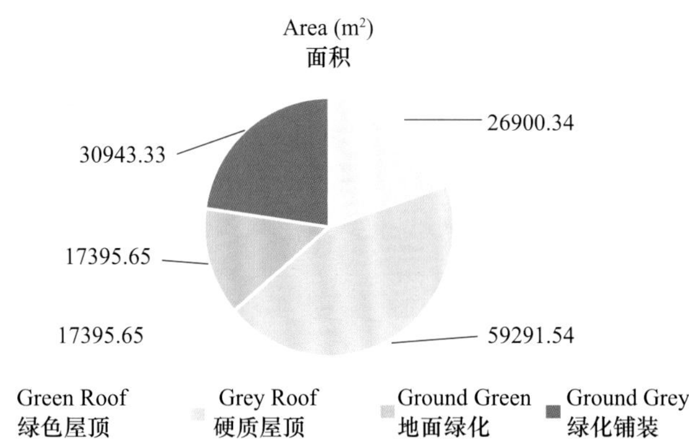

图 18 Pie chart of surface types and area in the site
场地表面类型与面积饼状图

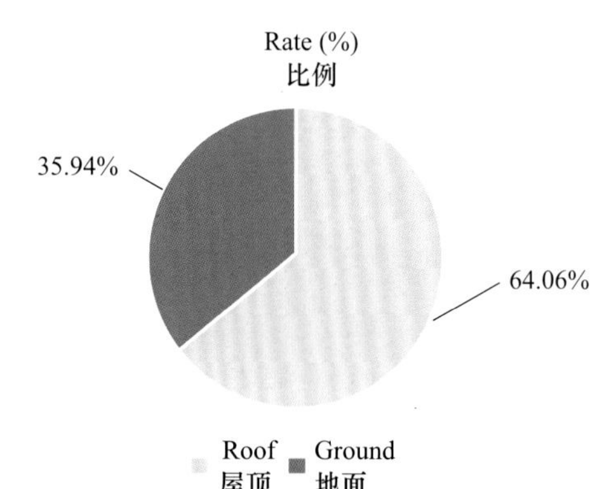

图 19 Pie chart of area ratio between roof and ground in the site
场地屋顶与地面面积比例饼状图

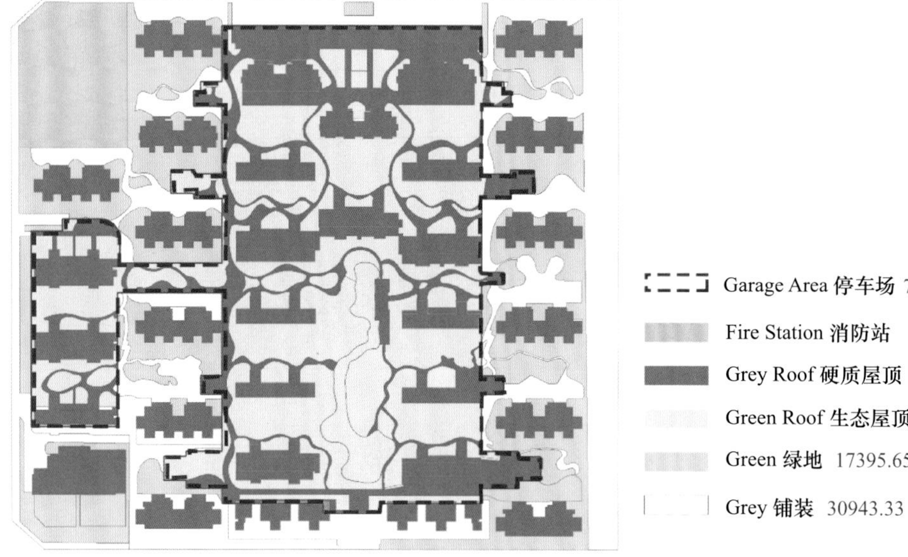

图 20 Roof and ground analysis
屋顶、地面分析

表 6 THE STATUS (UBC WAY) 现状（UBC 计算方法）——Extensive Roof 粗放型生态屋顶

	面积	Roof Area 屋顶面积	Extensive Rf 粗放型屋顶	Intensive Rf 集约型屋顶	Grey Roof 硬质屋顶		面积	Ground Area 地面面积	Green 绿地	Grey 硬质铺装
Roof 屋顶	Area	86191.88	26900.34	0	59291.54	**Ground** 地面	Area	48338.98	17395.65	30943.33
	CN	—	—	—	98		CN	—	61	98
	Kc	—	0.3	0.6	—					

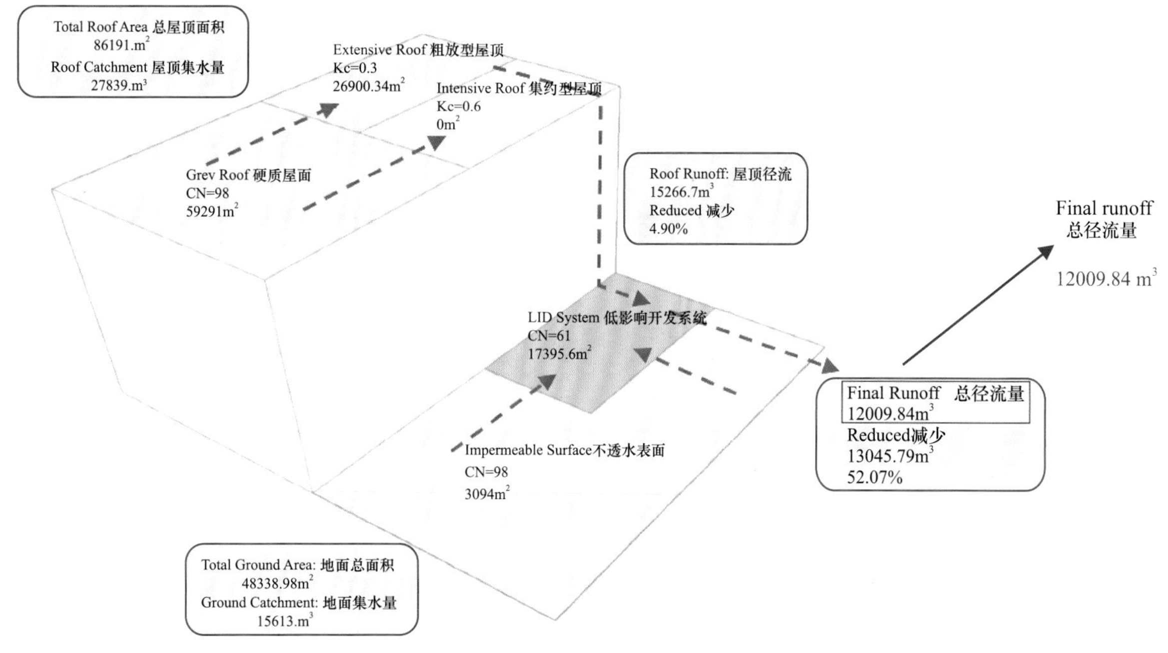

图 21　Runoff calculation of extensive sreen
粗放型生态屋顶径流量计算

表 7　THE STATUS (UBC WAY) 现状（UBC 计算法）——Intensive Roof 集约型生态屋顶

Roof 屋顶	面积	Roof Area 屋顶面积	Extensive Rf 粗放型屋顶	Intensive Rf 集约型屋顶	Grey Roof 硬质屋顶	Ground 地面	面积	Ground Area 地面面积	Green 绿地	Grey 硬质铺装
	Area	86191.88	0	26900.34	59291.54		Area	48338.98	17395.65	30943.33
	CN	—	—	—	98		CN	—	61	98
	Kc	—	0.3	0.6	—					

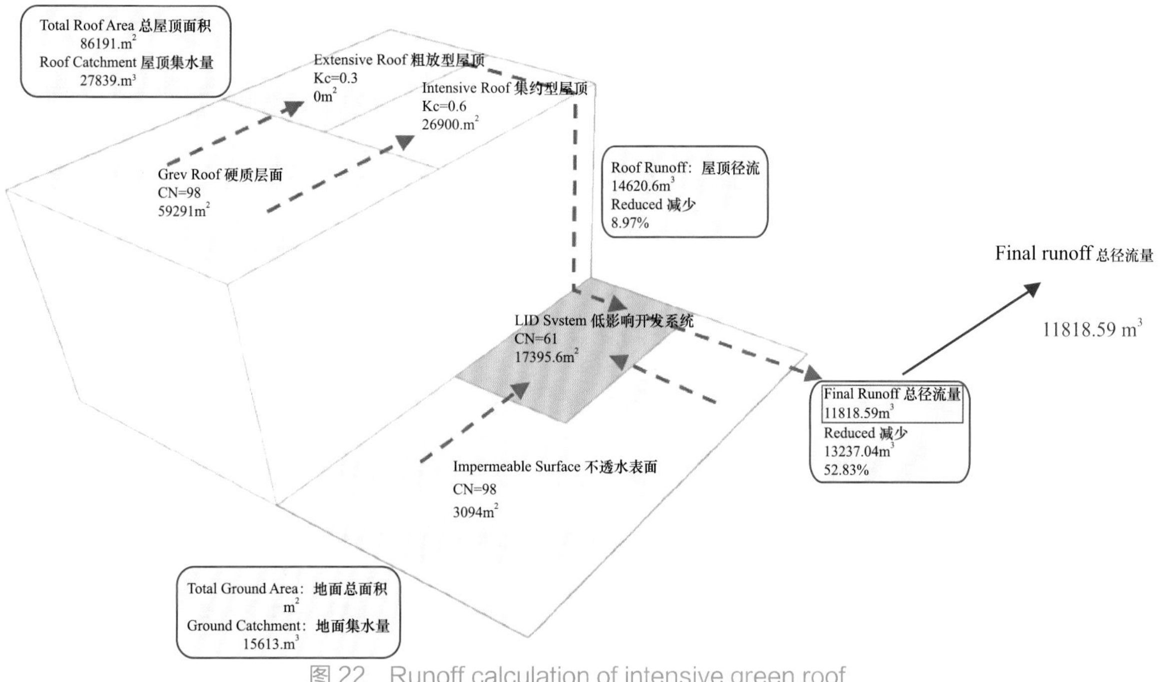

图 22　Runoff calculation of intensive green roof
集约型生态屋顶径流量计算

1.4 Pre-design calulation 设计前计算

1.4.1 Computational Thought and Hypothesis 计算思路与假设

2 ways:

Chinese way and UBC way

4 assumptions:

—change all possible grey roofs to green roofs with intensive plants

—change all the pavement to the permeable one

—change all the needless grey ground to green land

—combine these three

The reasons why we choose these three factors:

—We can calculate the area of the roofs which are possible to change into green roofs, the area of the pavement which are possible to change into permeable one, the area of the grey ground which we can change into green land based on the 15% standard.

—It is complex to calculate the area of ponds and tanks before details have been designed.

The aim we want to achieve by doing these job:

Form a rough concept of what we can do about the final runoff based on the current situation before we set the goal.

2 种方式：

中国方式和 UBC 方式

4 个假设：

将所有可能的灰色屋顶改为带有密集植物的绿色屋顶

将所有路面改为渗透性路面

将所有不必要的灰色地面改为绿地

将三者结合

我们选择这三个因素的原因：

我们可以计算出可以变成绿色屋顶的屋顶面积，可以变成渗透层的路面面积，我们可以根据 15%变成绿地的灰色地面面积 标准。

在设计细节之前计算池塘和水池的面积是很复杂的。

我们希望通过完成这些工作来实现目标：

在我们设定目标之前，根据当前的情况，形成一个关于最终径流的粗略概念。

1.4.2 Chinese calculation method 中式计算方法

ASSUMPTION 1（Chinese WAY）假设1（中式计算方法）

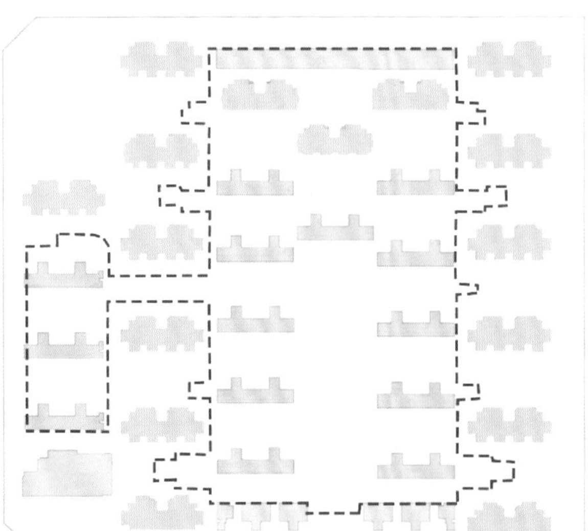

图 23 Transformation diagram of assumption1
假设 1 改造示意图

ASSUMPTION 1(Chinese WAY) 假设 1（中式计算方法）

表 8 CURRENT
现状

current 现状	area（m³）面积	runoff coefficient 径流系数		W（runoff/m³）总径流量	
type 类型				10-year period 10 年期	3-year period 3 年期
green roof 生态屋顶	0.00	0.30		0.00	0.00
road (impervious) 不透水路面	64163.57	0.90		12069.17	6236.70
green land 绿地	17396.65	0.15		545.38	281.83
grey roof 硬质屋面	9502.60	0.90		1787.44	923.65
basement which is covered with green land (green land above the garage) 绿化覆盖的地下室（停车场屋顶绿化）	26900.34	0.15		843.33	435.79
			sum	15245.32	7877.97

Final runoff 最终径流量
(10-year period) (10 年期)

15245.32m³

↓

16284.17m³

表 9 ASSUMPTION 1
假设 1

change all possible grey roof to green roof with intensive plants 带绿化覆盖的地下室和停车场	area（m³）面积	runoff coefficient 径流系数		W（runoff/m³）总径流量	
type 类型				10-year period 10 年期	3-year period 3 年期
green roof 生态屋顶	**16568.70**	**0.30**		1038.86	536.83
road (impervious) 不透水路面	64163.57	0.90		12069.17	6236.70
green land 绿地	17396.65	0.15		545.38	281.83
grey roof 硬质屋面	9502.60	0.90		1787.44	923.65
basement which is covered with green land (green land above the garage) 绿化覆盖的地下室（停车场屋顶绿化）	26900.34	0.15		843.33	435.79
			sum	16284.18	8414.80

ASSUMPTION 1 假设 1:
change all possible grey roof to green roof with intensive plants 将所有硬质屋顶换为集约型生态屋顶

By this extreme situation, we can reduce the final runoff by 11.32% of the current situation.
在这种极端情况下，我们可以比现状减少 11.32% 的径流量

ASSUMPTION 2(Chinese WAY) 假设 2（中式计算方法）

表 10 CURRENT
现状

current 现状	area（m³）面积	runoff coefficient 径流系数		W（runoff/m³）总径流量	
type 类型				10-year period 10 年期	3-year period 3 年期
green roof 生态屋顶	0.00	0.30		0.00	0.00
road (impervious) 不透水路面	64163.57	0.90		12069.17	6236.70
green land 绿地	17396.65	0.15		545.38	281.83
grey roof 硬质屋面	9502.60	0.90		1787.44	923.65
basement which is covered with green land (green land above the garage) 绿化覆盖的地下室（停车场屋顶绿化）	26900.34	0.15		843.33	435.79
			sum	15245.32	7877.97

Final runoff 最终径流量
(10-year period) (10 年期)

15245.32m³

↓

12144.44m³

表 11　ASSUMPTION 2
假设 2

change all pavement to the permeable one 将所有铺装换为透水铺装	area（m³）面积	runoff coefficient 径流系数		W（runoff/m³）总径流量	
type 类型				10-year period 10 年期	3-year period 3 年期
green roof 生态屋顶	0.00	0.30		0.00	0.00
road (impervious) 不透水路面	**64163.57**	**0.30**		5851.72	2078.90
green land 绿地	17396.65	0.15		545.38	281.83
grey roof 硬质屋面	26071.30	0.90		4904.01	2534.13
basement which is covered with green land (green land above the garage) 绿化覆盖的地下室（停车场屋顶绿化）	26900.34	0.15		843.33	435.79
			sum	12144.44	5330.65

ASSUMPTION 2 假设 2：
change all pavement to the permeable one 将所有铺装换为透水铺装

By this extreme situation, we can reduce the final runoff by **33.86%** of the current situation.
在这种极端情况下，我们可以比现状减少 33.86% 的径流量

ASSUMPTION 3(Chinese WAY) 假设 3（中式计算方法）

表 12　CURRENT
现状

current 现状	area（m³）面积	runoff coefficient 径流系数		W（runoff/m³）总径流量	
type 类型				10-year period 10 年期	3-year period 3 年期
green roof 生态屋顶	0.00	0.30		0.00	0.00
road (impervious) 不透水路面	64163.57	0.90		12069.17	6236.70
green land 绿地	17396.65	0.15		545.38	281.83
grey roof 硬质屋面	9502.60	0.90		1787.44	923.65
basement which is covered with green land (green land above the garage) 绿化覆盖的地下室（停车场屋顶绿化）	26900.34	0.15		843.33	435.79
			sum	15245.32	7877.91

Final runoff **最终径流量**
(10-year period) (10 年期)

15245.32m³
↓
8032.50m³

表 13　ASSUMPTION 3
假设 3

expand the green land to the extreme (retain 15% of the ground for road) 最大限度的扩展绿地（预留 15% 的地面作为道路）	area（m³）面积	runoff coefficient 径流系数		W（runoff/m³）总径流量	
type 类型				10-year period 10 年期	3-year period 3 年期
green roof 生态屋顶	0.00	0.30		0.00	0.00
road (impervious) 不透水路面	**14768.94**	**0.90**		2778.04	1435.54
green land 绿地	**83690.62**	**0.15**		2623.70	1355.79
grey roof 硬质屋面	9502.60	0.90		1787.44	923.65
basement which is covered with green land (green land above the garage) 绿化覆盖的地下室（停车场屋顶绿化）	26900.34	0.15		843.33	435.79
			sum	8032.51	4150.77

ASSUMPTION 3 假设 3：
expand the green land to the extreme (retain 15% of the ground for road) 最大限度的扩展绿地（预留 15% 的地面作为道路）

By this extreme situation, we can reduce the final runoff by **56.25%** of the current situation.
在这种极端情况下，我们可以比现状减少 56.25% 的径流量

ASSUMPTION 4(Chinese WAY) 假设 4（中式计算方法）

表 14　CURRENT
现状

current 现状 type 类型	area（m^3）面积	runoff coefficient 径流系数		W（runoff/m^3）总径流量 10-year period 10 年期	W（runoff/m^3）总径流量 3-year period 3 年期
green roof 生态屋顶	0.00	0.30		0.00	0.00
road (impervious) 不透水路面	64163.57	0.90		12069.17	6236.70
green land 绿地	17396.65	0.15		545.38	281.83
grey roof 硬质屋面	9502.60	0.90		1787.44	923.65
basement which is covered with green land (green land above the garage) 绿化覆盖的地下室（停车场屋顶绿化）	26900.34	0.15		843.33	435.79
			sum	15245.32	7877.97

Final runoff 最终径流量 (10-year period) (10 年期)

18361.89m^3

6180.48m^3

表 15　ASSUMPTION 4
假设 4

Combine all the extreme situations 结合上述所有极端情况 type 类型	area（m^3）面积	runoff coefficient 径流系数		W（runoff/m^3）总径流量 10-year period 10 年期	W（runoff/m^3）总径流量 3-year period 3 年期
green roof 生态屋顶	**16568.70**	**0.30**		0.00	0.00
road (impervious) 不透水路面	**14768.94**	**0.30**		926.01	478.51
green land 绿地	**83690.62**	**0.15**		2623.70	1355.79
grey roof 硬质屋面	**9502.60**	**0.90**		1787.44	923.65
basement which is covered with green land (green land above the garage) 绿化覆盖的地下室（停车场屋顶绿化）	26900.34	0.15		843.33	435.79
			sum	6180.48	3193.74

ASSUMPTION 4 假设 4：
Combine all the extreme situations. 结合上述所有极端情况

By this extreme situation, we can reduce the final runoff by **66.34%** of the current situation.
在这种极端情况下，我们可以比现状减少 66.34% 的径流量

This is the maximum amount of runoff we can reduce under different situations considering 10 year period, but this is the ideal situation.
这是不同情况下对十年一遇的雨洪我们能减少径流量最多的一种方法了，但它只是理想值

This is the maximum amount of runoff we can reduce under different situations considering 3 year Event, but this is the ideal situation.
这是不同情况下对三年一遇的雨洪我们能减少径流量最多的一种方法了，但它只是理想值

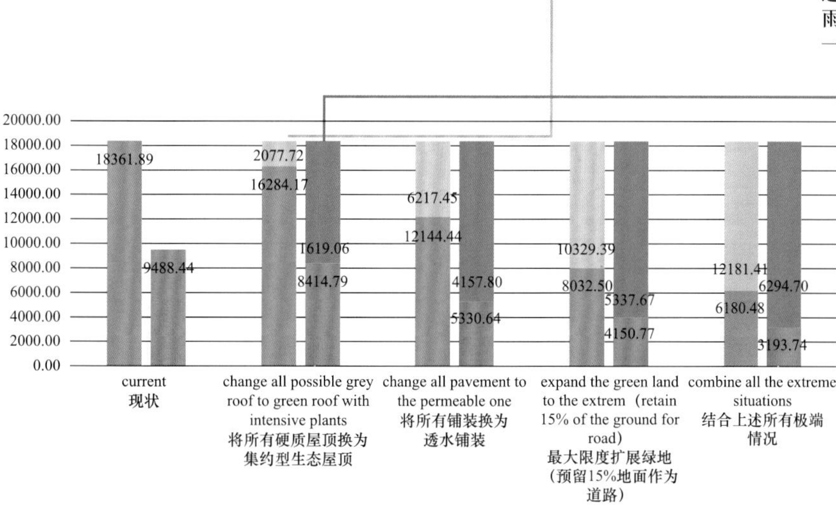

10 year period (worse than the3 year period)
十年一遇情况下的减少比例（十年一遇比三年一遇情况更严重）

图 24　Comparison (chinese way)
对比（中式方法）

1.4.3 UBC's Calculation method UBC 计算方法

ASSUMPTION1:
Change all the roofs of high-rise residential buildings togreen roofs with intensive plants.

假设1:
将所有高层住宅楼屋顶改为集约型生态屋顶

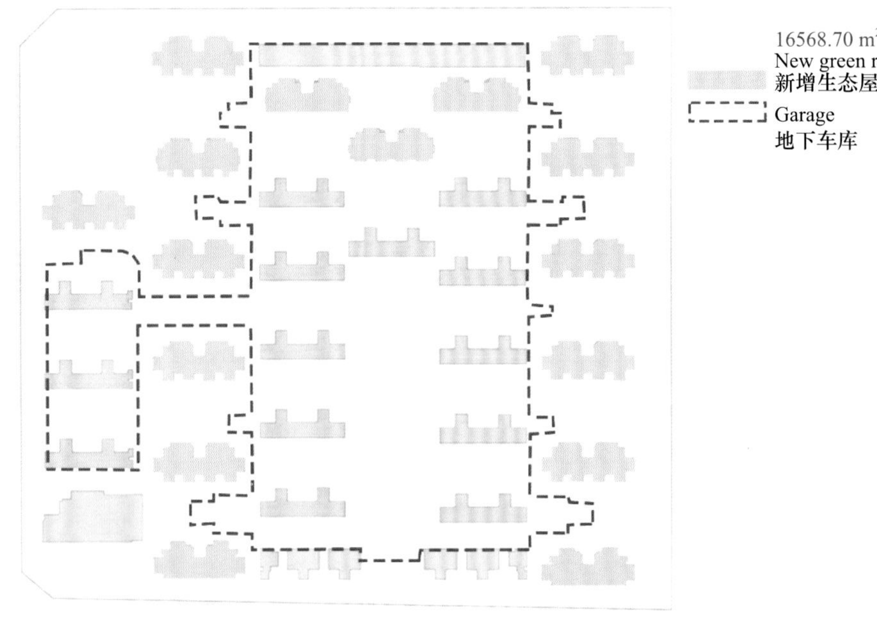

图 25 Assumption 1(UBC WAY)
假设 1（UBC 计算方法）

ASSUMPTION 1 (UBC WAY) 假设 1（UBC 方法）

表 16 CURRENT
现状

	面积	Roof Area 屋顶面积	Extensive Rf 粗放型生态屋顶	Intensive Rf 集约型生态屋顶	Grey Roof 硬质屋顶		面积	Ground Area 地面面积	Green 绿地	Grey 硬质铺装	Final runoff 最终径流量
Roof 屋顶	Area	86191.88	26900.34	0	59291.54	Ground 地面	Area	48338.98	17395.65	30943.33	12009.84m^3
	CN	—	—	—	98		CN	—	61	98	
	Kc	—	0.3	0.6	—						

表 17 ASSUMPTION 1
假设 1

	面积	Roof Area 屋顶面积	Extensive Rf 粗放型生态屋顶	Intensive Rf 集约型生态屋顶	Grey Roof 硬质屋顶		面积	Ground Area 地面面积	Green 绿地	Grey 硬质铺装	Final runoff 最终径流量
Roof 屋顶	Area	86191.88	0	43469.04	59291.54	Ground 地面	Area	48338.98	17395.65	30943.33	11540.79m^3
	CN	—	—	—	98		CN	—	61	98	
	Kc	—	0.3	0.6	—						

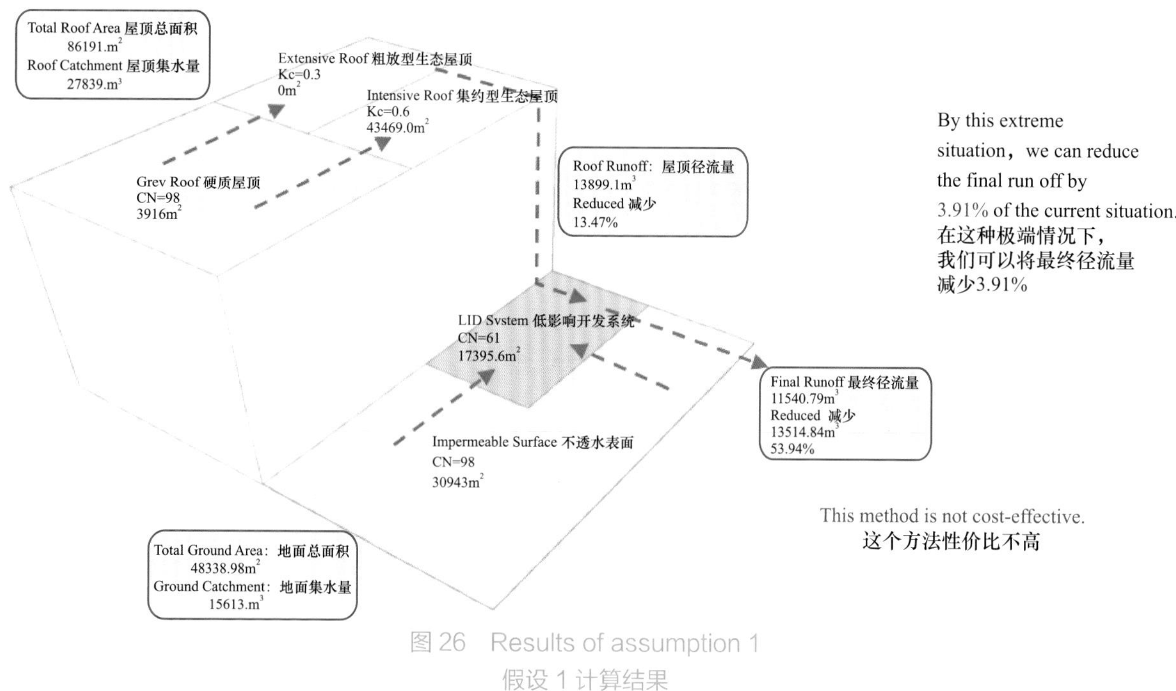

图 26 Results of assumption 1
假设 1 计算结果

ASSUMPTION 2 (UBC WAY) 假设 2（UBC 方法）

表 18 CURRENT
现状

Roof 屋顶	面积	Roof Area 屋顶面积	Extensive Rf 粗放型生态屋顶	Intensive Rf 集约型生态屋顶	Grey Roof 硬质屋顶	Ground 地面	面积	Ground Area 地面面积	Green 绿地	Grey 硬质铺装	Final runoff 最终径流量
	Area	86191.88	26900.34	0	59291.54		Area	48338.98	17395.65	30943.33	
	CN	—	—	—	98		CN	—	61	98	
	Kc	—	0.3	0.6	—						

12009.84m^3 → 7643.02m^3

表 19 ASSUMPTION 2
假设 2

Roof 屋顶	面积	Roof Area 屋顶面积	Extensive Rf 粗放型生态屋顶	Intensive Rf 集约型生态屋顶	Grey Roof 硬质屋顶	Ground 地面	面积	Ground Area 地面面积	Green 绿地	Grey 硬质铺装
	Area	86191.88	0	43469.04	59291.54		Area	48338.98	17395.65	30943.33
	CN	—	—	—	98		CN	—	61	65
	Kc	—	0.3	0.6	—					

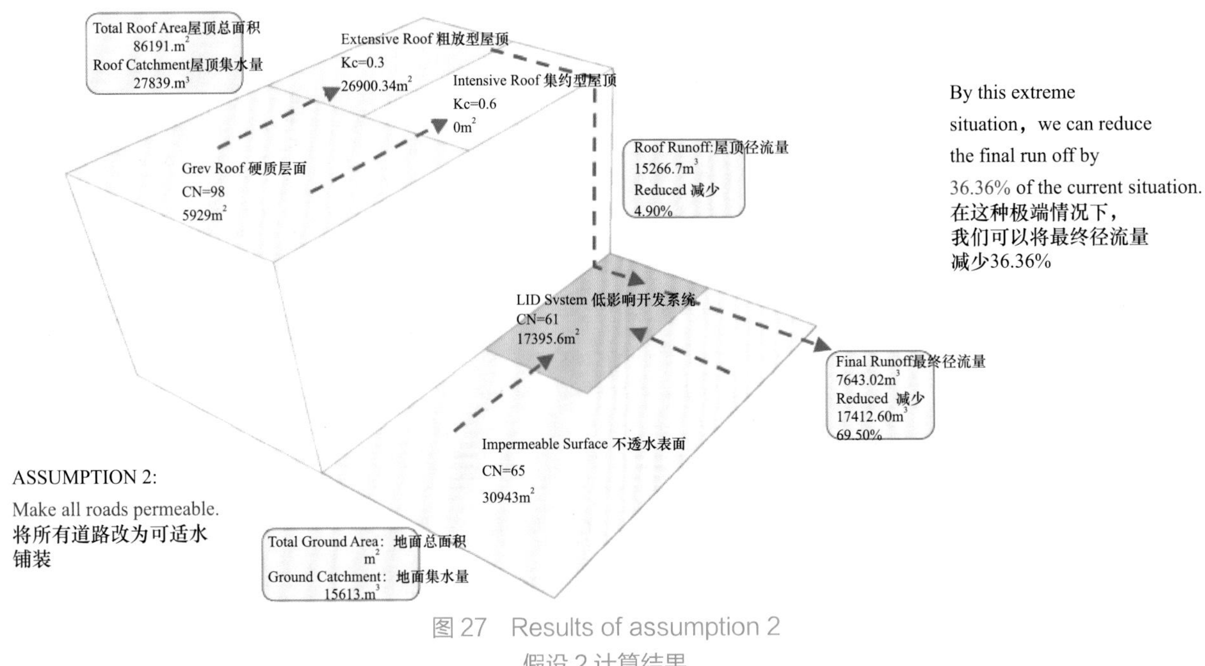

ASSUMPTION 2:

Make all roads permeable.
将所有道路改为可透水铺装

By this extreme situation, we can reduce the final run off by 36.36% of the current situation.
在这种极端情况下，我们可以将最终径流量减少36.36%

图 27 Results of assumption 2
假设 2 计算结果

ASSUMPTION 3 (UBC WAY) 假设 3（UBC 方法）

表 20 CURRENT
现状

Roof 屋顶	面积	Roof Area 屋顶面积	Extensive Rf 粗放型生态屋顶	Intensive Rf 集约型生态屋顶	Grey Roof 硬质屋顶	Ground 地面	面积	Ground Area 地面面积	Green 绿地	Grey 硬质铺装	Final runoff 最终径流量
	Area	86191.88	26900.34	0	59291.54		Area	48338.98	17395.65	30943.33	
	CN	—	—	—	98		CN	—	61	98	
	Kc	—	0.3	0.6	—						12009.84m³

表 21 ASSUMPTION 1
假设 1

Roof 屋顶	面积	Roof Area 屋顶面积	Extensive Rf 粗放型生态屋顶	Intensive Rf 集约型生态屋顶	Grey Roof 硬质屋顶	Ground 地面	面积	Ground Area 地面面积	Green 绿地	Grey 硬质铺装	Final runoff 最终径流量
	Area	86191.88	69502.83	0	16689.05		Area	48338.98	41088.13	7250.85	5840.93m³
	CN	—	—	—	98		CN	—	61	98	
	Kc	—	0.3	0.6	—						

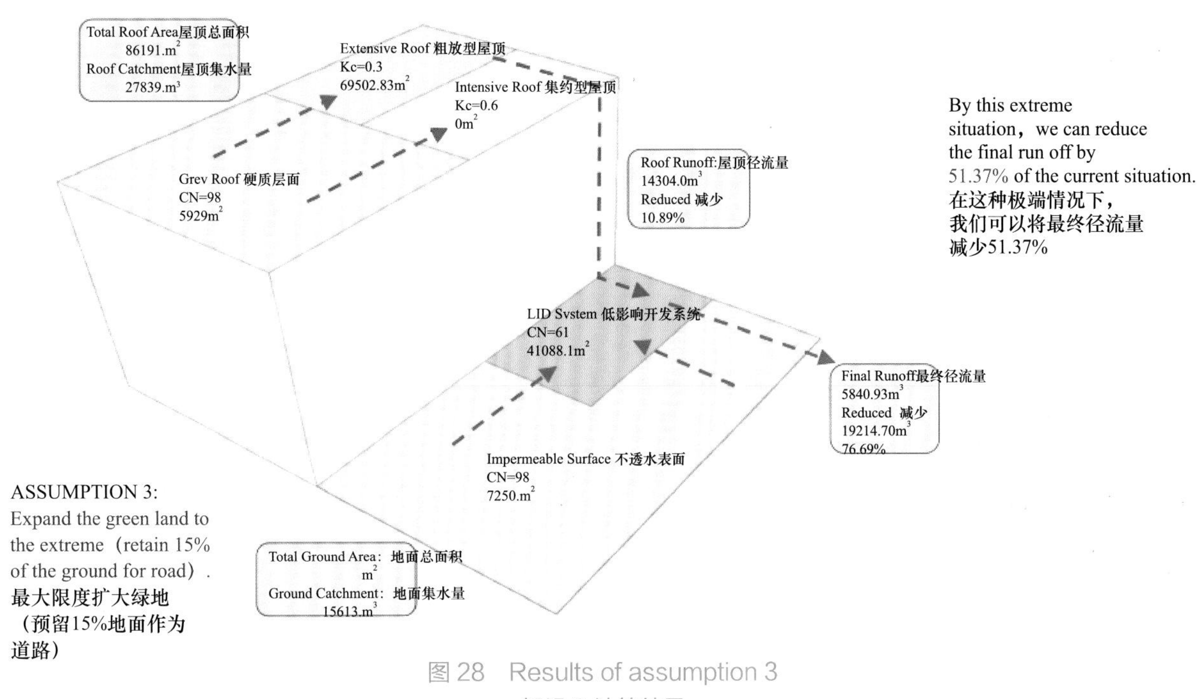

By this extreme situation，we can reduce the final run off by 51.37% of the current situation.
在这种极端情况下，我们可以将最终径流量减少51.37%

ASSUMPTION 3:
Expand the green land to the extreme（retain 15% of the ground for road）.
最大限度扩大绿地（预留15%地面作为道路）

图 28 Results of assumption 3
假设 3 计算结果

ASSUMPTION 3 (UBC WAY) 假设 3（UBC 方法）

表 22 CURRENT
现状

	面积	Roof Area 屋顶面积	Extensive Rf 粗放型生态屋顶	Intensive Rf 集约型生态屋顶	Grey Roof 硬质屋顶		面积	Ground Area 地面面积	Green 绿化	Grey 铺装	Final runoff 最终径流量
Roof 屋顶	Area	86191.88	26900.34	0	59291.54	Ground 地面	Area	48338.98	17395.65	30943.33	
	CN	—	—	—	98		CN	—	61	98	
	Kc	—	0.3	0.6	-						12009.84m³

↓

2984.19m³

表 23 ASSUMPTION 1
假设 1

	面积	Roof Area 屋顶面积	Extensive Rf 粗放型生态屋顶	Intensive Rf 集约型生态屋顶	Grey Roof 硬质屋顶		面积	Ground Area 地面面积	Green 绿化	Grey 铺装
Roof 屋顶	Area	86191.88	60271.19	0	17920.69	Ground 地面	Area	48338.98	41088.13	7250.85
	CN	—	—	—	98		CN	—	61	65
	Kc	—	0.3	0.6	—					

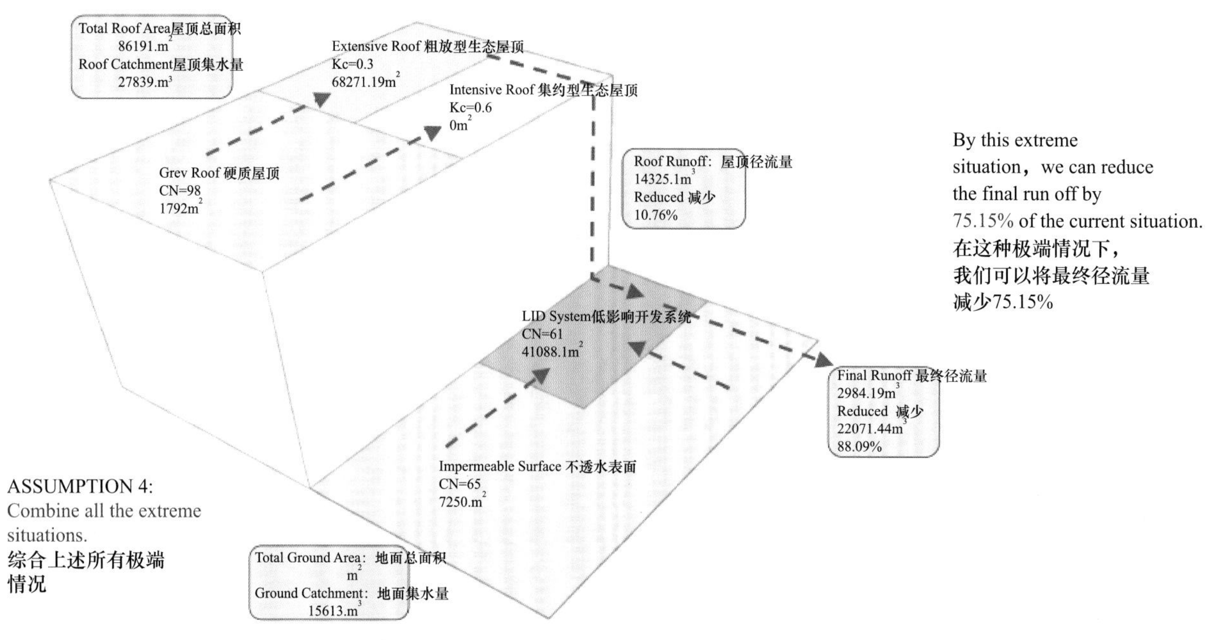

图 29 Results of assumption 4

假设 4 计算结果

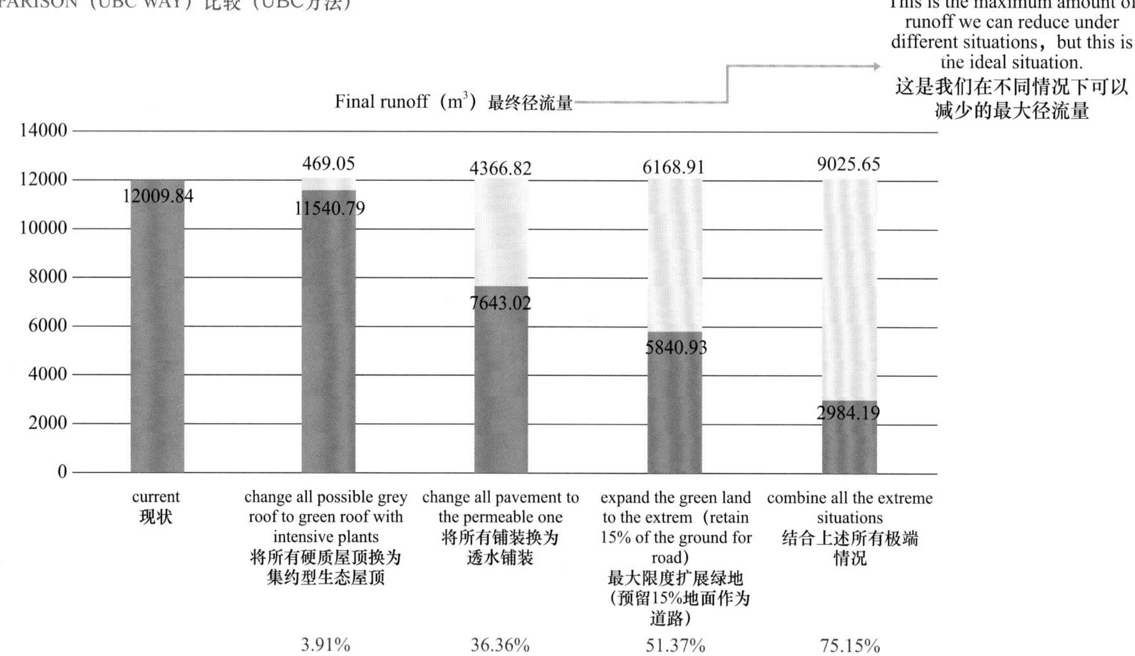

图 30 Comparison of final runoff among current situation and four assumptions

现状与四种假设总径流量比较

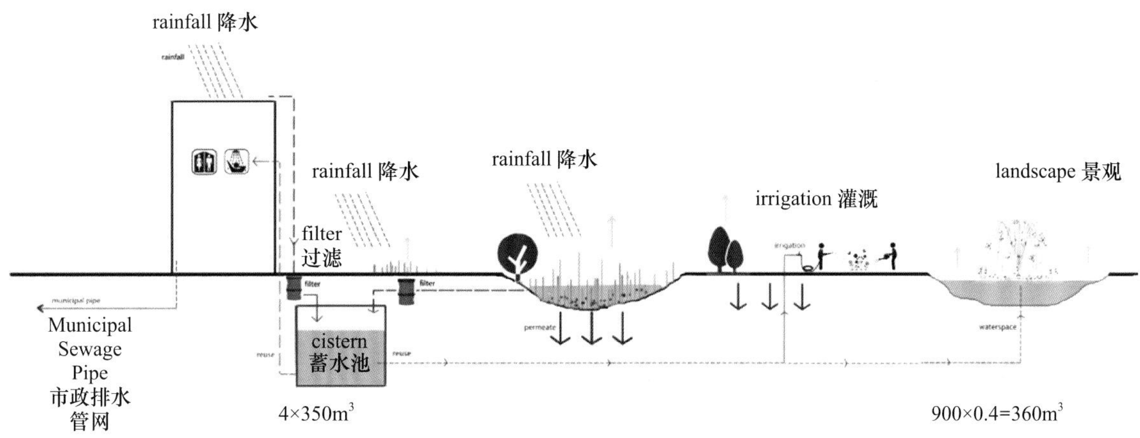

图 31 Current water flow diagran
现状雨水径流图

2. LID in Gaobeidian 高碑店 LID 项目学习

图 32 WATER RECHARGE
水再利用
Use collected rainwater for irrigation.
用收集的雨水灌溉

图 33 GREENROOF
(intensive)
集约型生态屋顶

图 34 WATER PURIFICATION
水净化设施
Use microorganisms，animals and plants to purify water.
利用微生物、动物、植物来净水

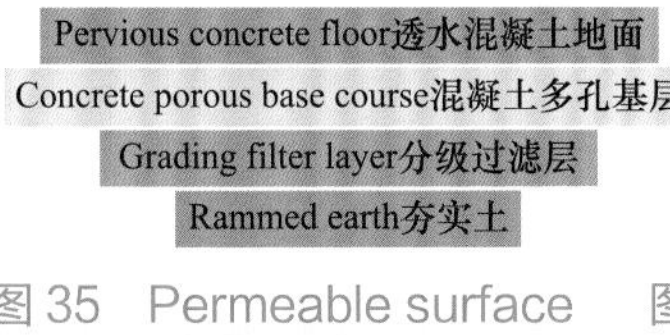

图 35 Permeable surface
透水地面

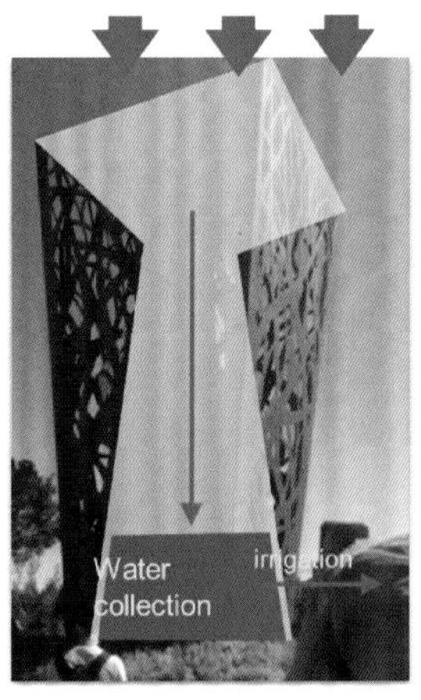

图 36 Rainwater collect tower
雨水收集塔

图 37 Rainwater garden
雨水花园

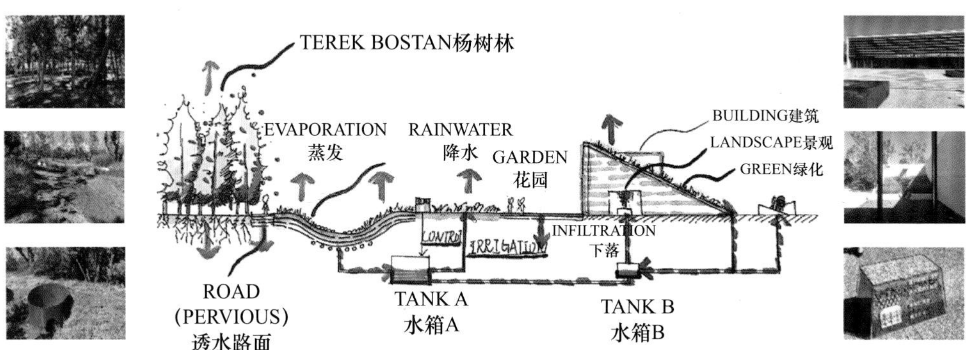

图 38 Gaobeidian LID measures section
高碑店 LID 策略剖面

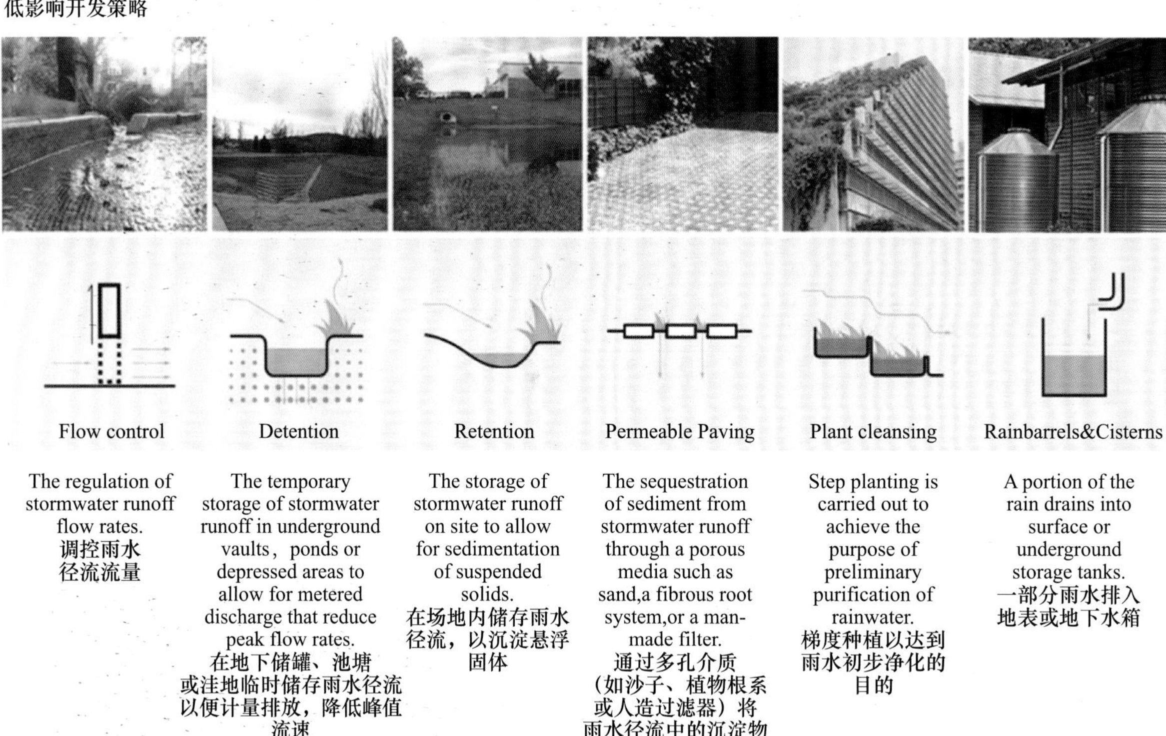

图 39 LID measures
LID 策略总结

3. Design Results 设计成果

3.1 Design results of group one 第一组设计成果

3.1.1 Design method and system construction 设计方法与系统构建

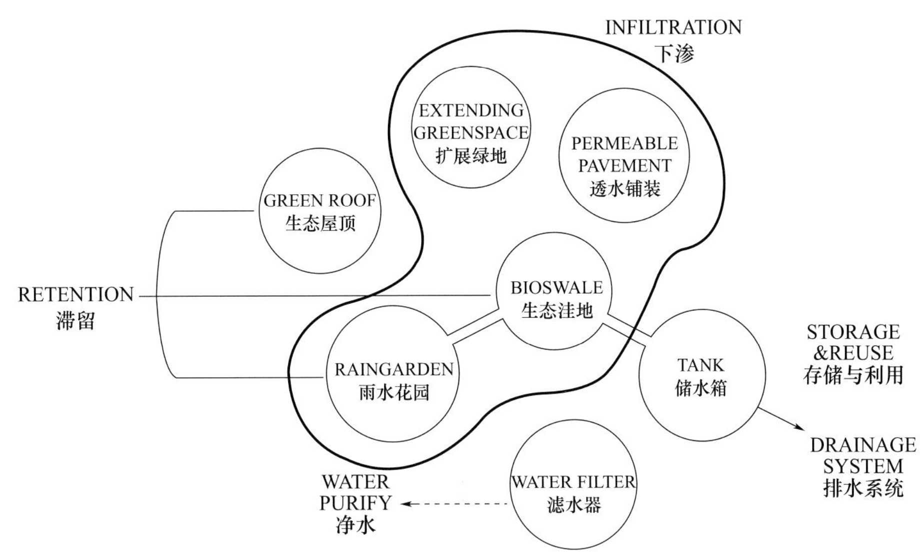

图 40 Design approaches
设计方法

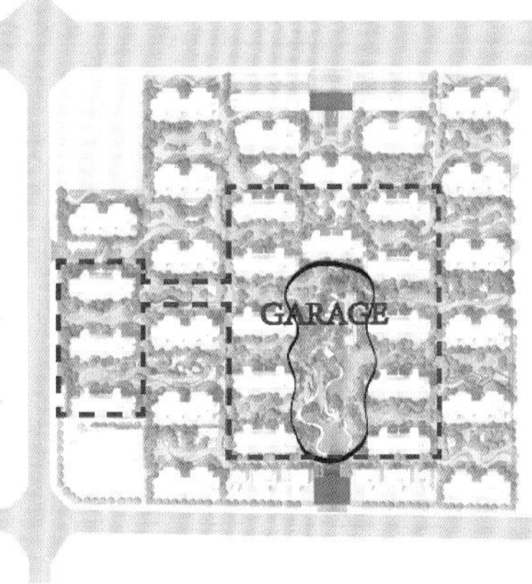

Within the garage:
停车场内部

- Green space&roads→bioswale→pond
- 绿地与道路→生态洼地→水池
- Grey roof→tank→bioswale→rain garden→drainage system
- 硬质屋顶→水箱→生态洼地→雨水花园→排水系统

Outside the garage:
停车场外部

- Grey roof→tank→bioswale→rain garden→drainage system
- 硬质屋顶→水箱→生态洼地→雨水花园→排水系统
- Green roof→Water filter→tank→bioswale→rain garden→drainage system
- 生态屋顶→净水器→水箱→生态洼地→雨水花园→排水系统

图 41 System construction
系统构建

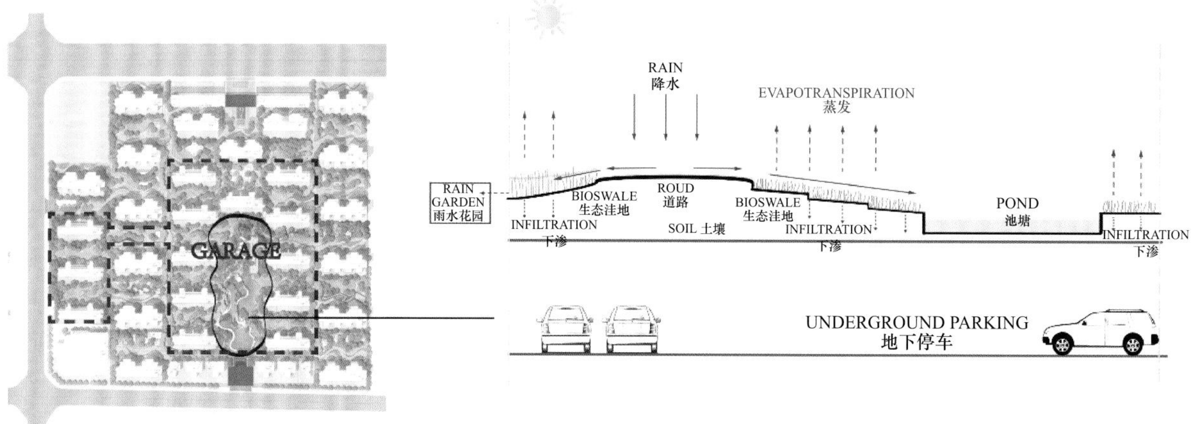

图 42　System1 within the garage—from green space to the pond
停车场系统 1——从绿地到水池

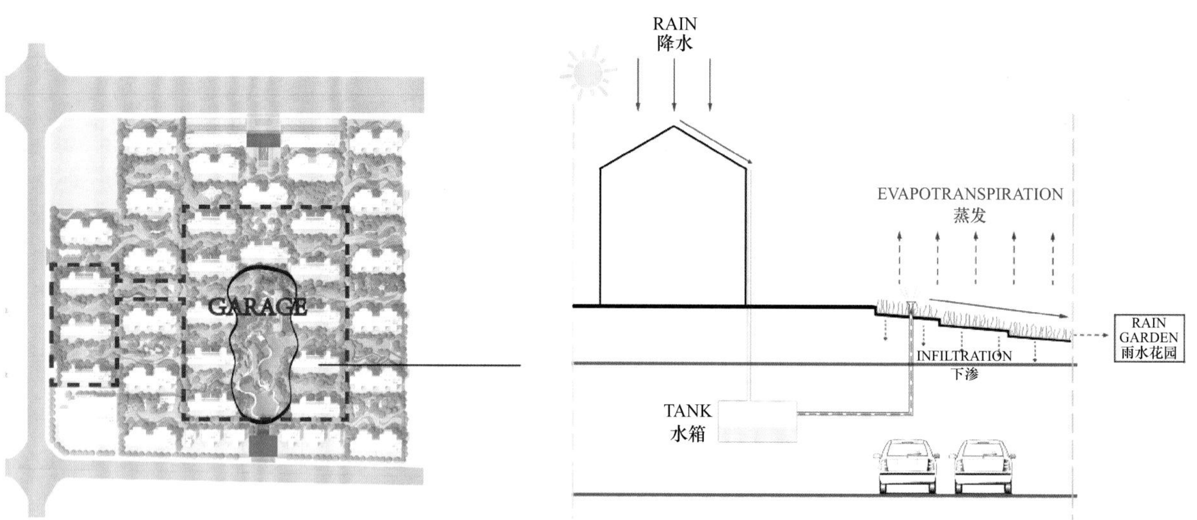

图 43　System2 within the garage—from grey roof to green roof
停车场系统 2——从普通屋顶到绿色屋顶

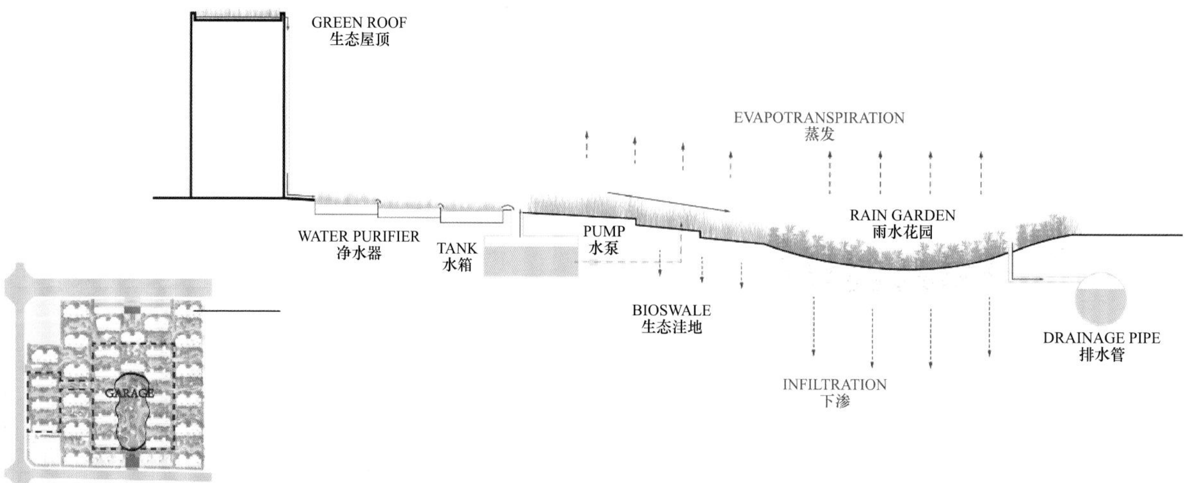

图 44　System3 water management of green roof
系统 3 绿色屋顶雨水管理

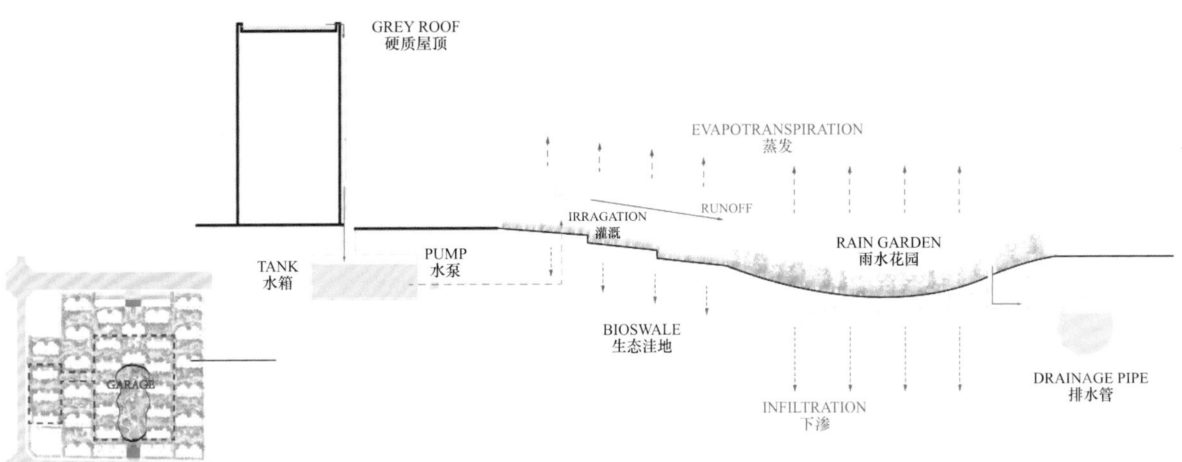

图 45　System4 water management of grey roof
系统 4 普通屋顶雨水管理

3.1.2 detail design 细节设计

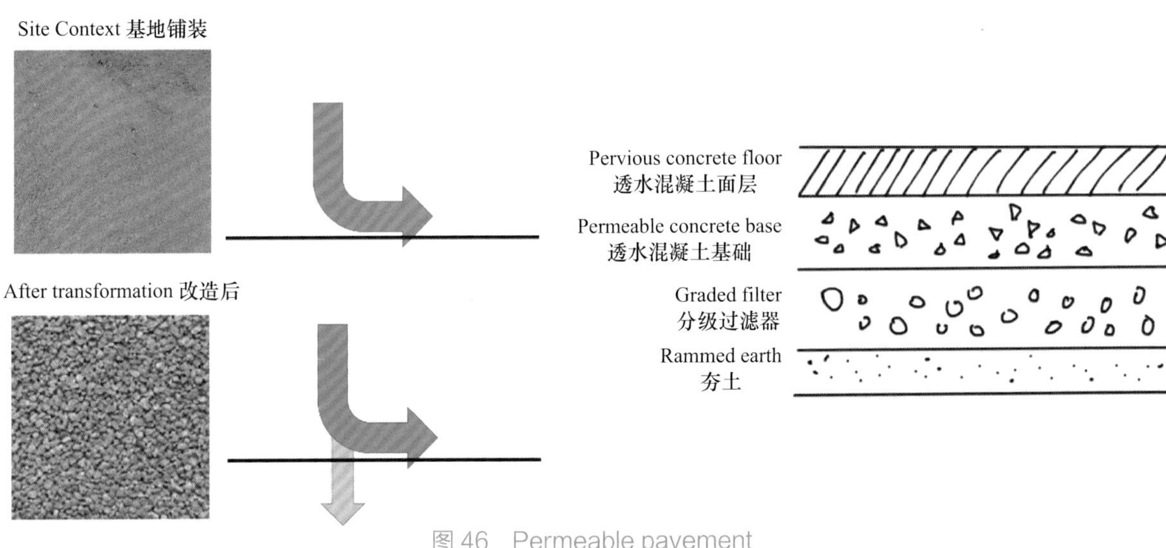

图 46　Permeable pavement
透水铺装

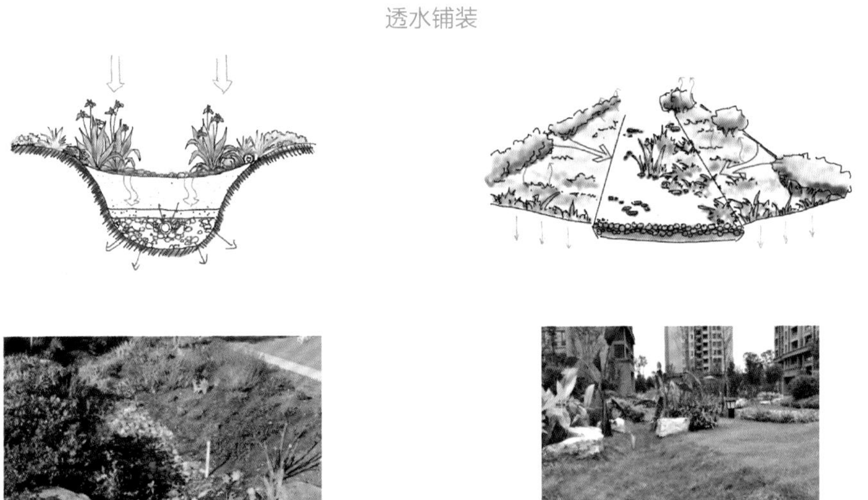

图 47　Rain garden and bioswale
雨水花园与生态洼地

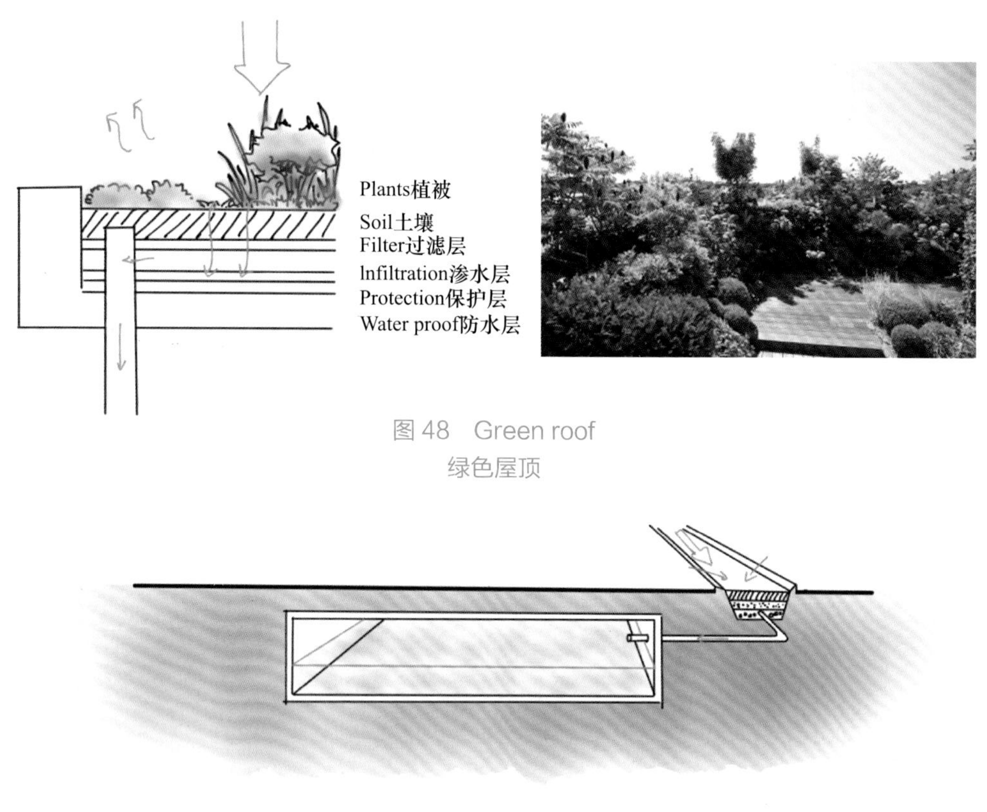

图 48　Green roof
绿色屋顶

图 49　Tank
储水罐

Clean water is filtered through the rain garden, bioswale and roof greening and into the cistern.

通过雨水花园、生态洼地和屋顶绿化的过滤，清洁的水流入水箱中。

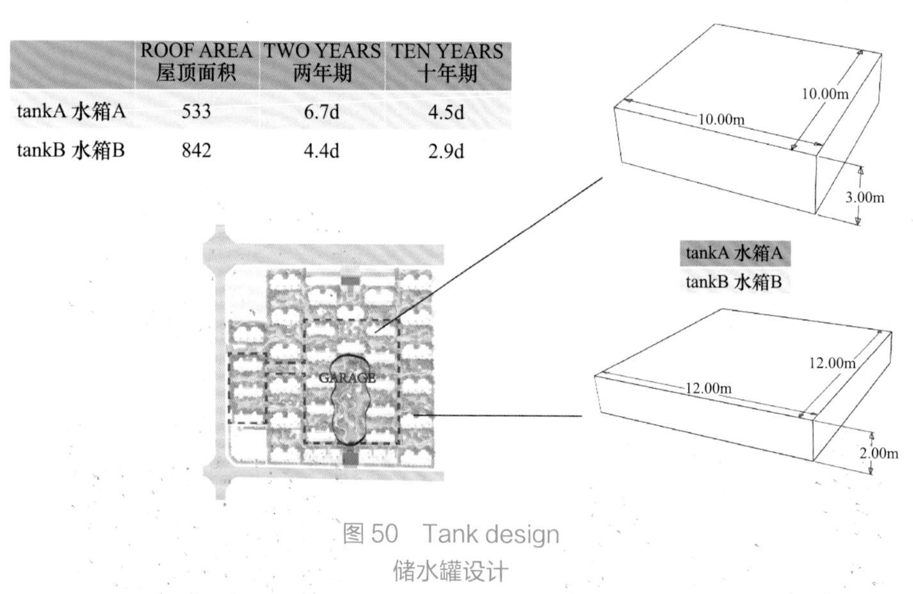

	ROOF AREA 屋顶面积	TWO YEARS 两年期	TEN YEARS 十年期
tankA 水箱A	533	6.7d	4.5d
tankB 水箱B	842	4.4d	2.9d

图 50　Tank design
储水罐设计

图 51 Detailed plan example 1
详细设计 1

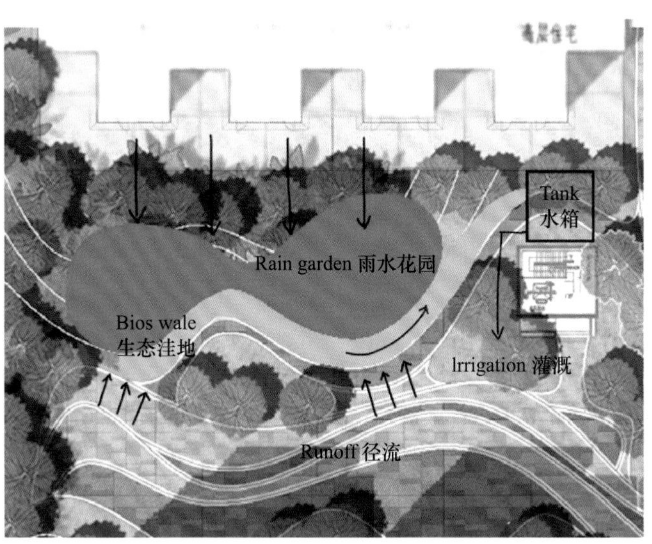

图 52 Detailed plan example 2
详细设计 2

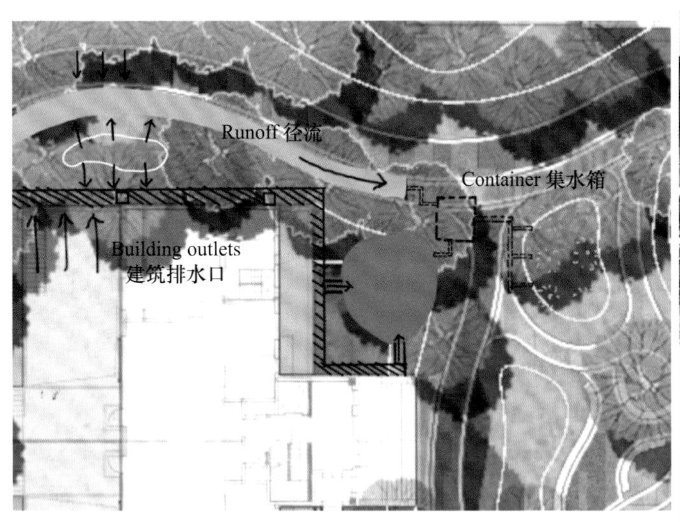

图 53 Detailed plan example 3
详细设计 3

3.1.3 Planting design 植物设计

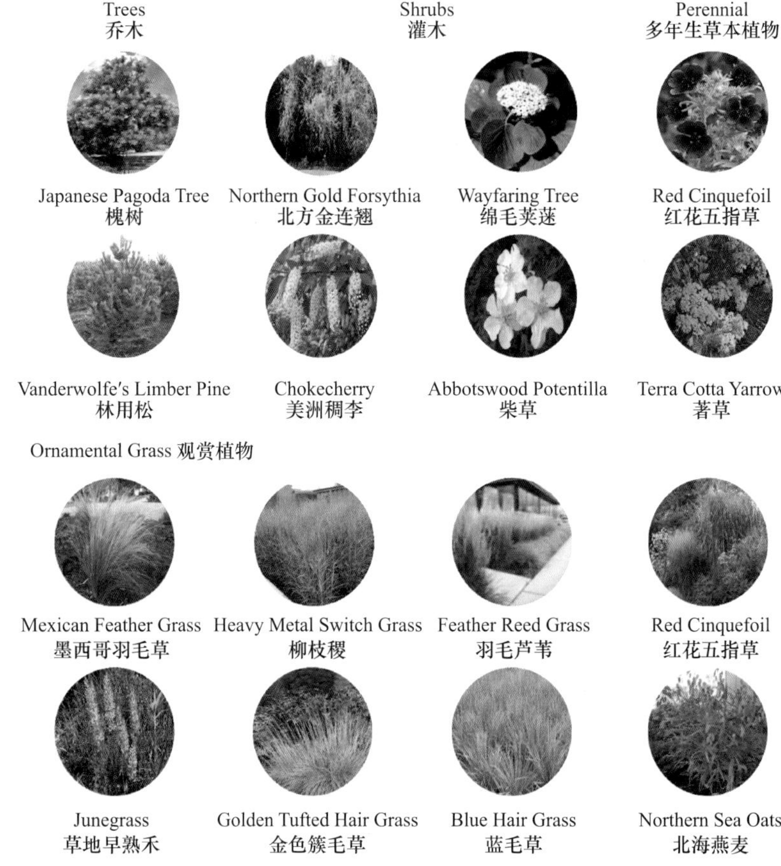

EXTENSIVE GREEN ROOF
粗放型生态屋顶

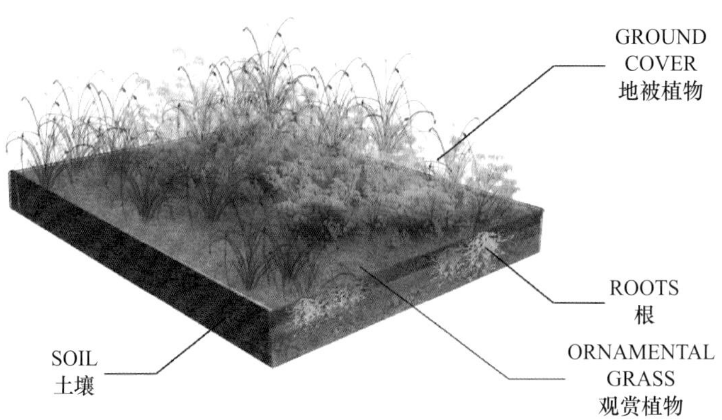

Extensive Green Roof Plant Schedule
粗放型生态屋顶种植设计

	Common Name 常用名	Latin Name 拉丁名	Type 类别	Water consumption 耗水量	Height 高度	Spreed 生长半径
	Angelina Stanecrop 岩景天	Sedum rupestre *'Angclina'*	Ground cover& Perennial 地被植物/多年生植物	💧	18-24′	18-24′
	Autumn Flre Sedum 秋火景天	Sedum *'Autumn Fire'*	Perennial 多年生植物	💧	4-6′	24′
	White Stonecrop 白景天	*Sedum aibum*	Ground cover& Perennial 地被植物/多年生植物	💧	2-4′	1-2′
	Alyssum, Sweet Alyssum 香雪球	*Lobularla maritima*	Annual 一年生植物	💧	4′	6-12″
	Elijah Blue Fescue 以利亚蓝羊茅	*Festuca glauea'Eiijaf Blue'*	Ground cover& Perennial grass 地被植物/多年生植物	💧	12″	10-15″
	Idaho Fescue 爱达荷羊茅	*Festuca idahoensis*	Ground cover& Perennial grass 地被植物/多年生植物	💧💧	12″	12″
	Papnka Yarrow 辣椒	*Achlllea mjllefoiium 'Paprika'*	Perennial 多年生植物	💧💧	18-24″	24″
	Terra Cotta Yarrow 赤辣椒	*Achlllea mjllefoiium 'Terra Cotta'*	Perennial 多年生植物	💧💧	30-36″	24″
	Abbotswood Potentilla 柴草	*Potentilla frulticosa 'Abbottswood'*	Shrub 灌木	💧💧	2′	3′
	Red Clnquefoll 红桂皮	*Potentilla thurberi 'Monarch's Velvel'*	Perennial 多年生植物	💧💧	12′	12′

INTENSIVE GREEN ROOF
集约型生态屋顶

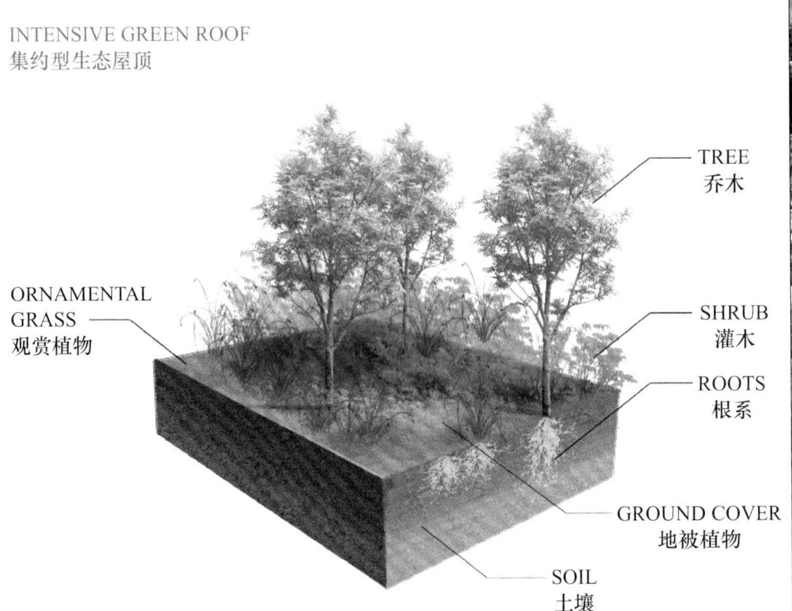

Intensive Green Roof Plant Schedule 集约型生态屋顶种植设计

	Common Name 常用名	Latin Name 拉丁名	Type 类别	Water consumption 耗水量	Height 高度	Spreed 生长半径
	Japanese Pagoda Tree 槐树	*Sophora japonica*	Tree 乔木		40-60′	30-45′
	Northern Gold Forsythia 北方金连翘	*Forsythra Northem Gold'*	Shrub 灌木		6-10′	6-12′
	Chokecherry 北美稠李	*Prunus virginiana*	Shrub 灌木		12′	6′
	Waylaring Tree 绵毛荚蒾	*Viburnum lantana*	Shrub& Tree 灌木或小乔木		10-15′	10-15′
	Vanderwolfe's Limber Pine 美国果松	*Pinus Fcxis 'Vanoerwoffe' 's'*	Tree 乔木		26′	16′
	Idaho Fescue 爱达荷羊茅	*Festuca idahoensis*	Ground cover& Ornamental grass 地被植物或观赏草		12″	12″
	Paprika Yarrow 辣椒	*Achillea millefolium 'Paprika'*	Perennial 多年生植物		18-24″	24″
	Terra Cotta Yarrow 赤辣椒	*Achillea millefolium 'Terra Cotta'*	Perennial 多年生植物		30-36″	24″
	Abbotswood Potentilla 柴草	*Potentilla fruiticosa 'Abbottswood'*	Shrub 灌木		2′	3′
	Red Cinquefoil 红桂皮	*Potentilla fruiticosa 'Monareh's Velvet'*	Perennial 多年生植物		12′	12′

RAIN GARDEN
雨水花园

Rain Garden Plant Schedule 雨水花园种植设计

	Common Name 常用名	Latin Name 拉丁名	Type 类别	Water consumption 耗水量	Height 高度	Spreed 生长半径
	Nartharn Sea Oats 北海燕麦	*Chesmenthium labfollum*	Omamental grass 观赏草		36″	24″
	Elijah Blue Fascua 以利亚蓝羊茅	*Festuca glauca 'Eiijah ekc'*	Ground cover& Ornamental grass 地被植物或观赏草		12″	10-15″
	German Iris 德国鸢尾	*Iyis germanica*	Perennial 多年生植物		4-40″	6-24′
	Silvcr Variegated Sweet Flag(Iris) 香根鸢尾	*Ins pallida 'Argenba-Varlegata'*	Perennial 多年生植物		24″	18″
	Feather Reed Grass 羽毛芦苇	*Calarnagrostis 'Kan Foerster*	Omamental grass 观赏草		60″	36″
	Japanese Pagada Tree 槐树	*Sophora japonico*	Tree 乔木		40-60′	30-45′
	Northem Gold Farsythla 北方金连翘	*Forsythia 'Northem Gold'*	Shrub 灌木		6-10′	6-12′
	Chokecherry 北美稠李	*Prunus virginiena*	Shrub 灌木		12′	6′
	Vanderwolfe's Limber Pine 美国果松	*Pinus Bexis 'Vander noite's'*	Tree 乔木		26′	16′
	Globe Blue Spruce 蓝叶云杉	*Pices pungens 'Globosa'*	Shrub 灌木		4′	7′

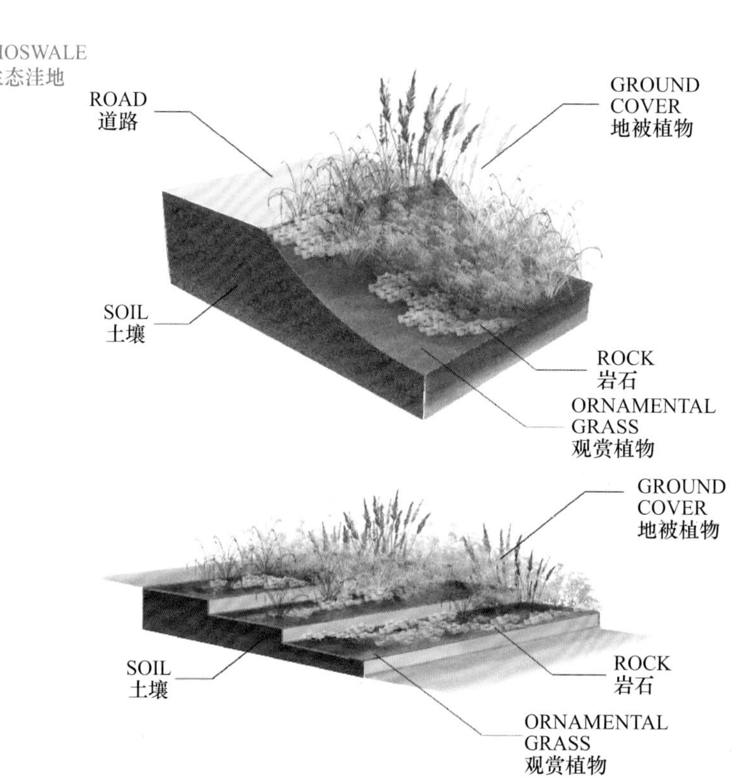

Bio Swale Plant Schedule
生态洼地种植设计

	Common Name 常用名	Latin Name 拉丁名	Type 类别	Water consumption 耗水量	Height 高度	Spreed 生长半径
	Giant Chinese Silver Grass 芒草	*Miscanthus 'Glganteus'*	Omamental grass 观赏草	💧💧💧	40-60′	30-45′
	Elijah Blue Fascua 以利亚蓝羊茅	*Festuca glauca 'Elijah Blue'*	Ground cover& Ornamental grass 地被植物或观赏草	💧	12″	10-15″
	Idaho Fescue 爱达荷羊茅	*Festuca ldahoensfs*	Ground cover& Ornamental grass 地被植物或观赏草	💧💧	12″	12″
	Blue Hair Grass 绵毛荚蒾	*Viburnum fantana*	Omamental grass 观赏草	💧	12″	12″
	Feather Reed Grass 羽毛芦苇	*Calemagrostis 'Kan Foerster'*	Omamental grass 观赏草	💧💧💧	60″	36″
	Golden Tufted Hair Grass 金色簇毛草	*Deschampsia fiexuosa 'Aurea'*	Omamental grass 观赏草	💧	8-18″	12″
	Heavy Metal Switch Grass 柳枝稷	*Panicum virgetum 'Heavy Metal'*	Omamental grass 观赏草	💧💧	54″	36″
	Junegrass 草地早熟禾	*Koeleria cristata (macrantha)*	Omamental grass 观赏草	💧	18″	12″
	Mexican Feather Grass 墨西哥针茅	*Stipa tenuissima*	Omamental grass 观赏草	💧	12-18″	12-18″
	Variegated Japanese Sedge 杂色日本莎草	*Carex oshimensis 'Evergold'*	Omamental grass 观赏草	💧💧💧	12″	18″

3.1.4 Design result accounting 设计成果核算

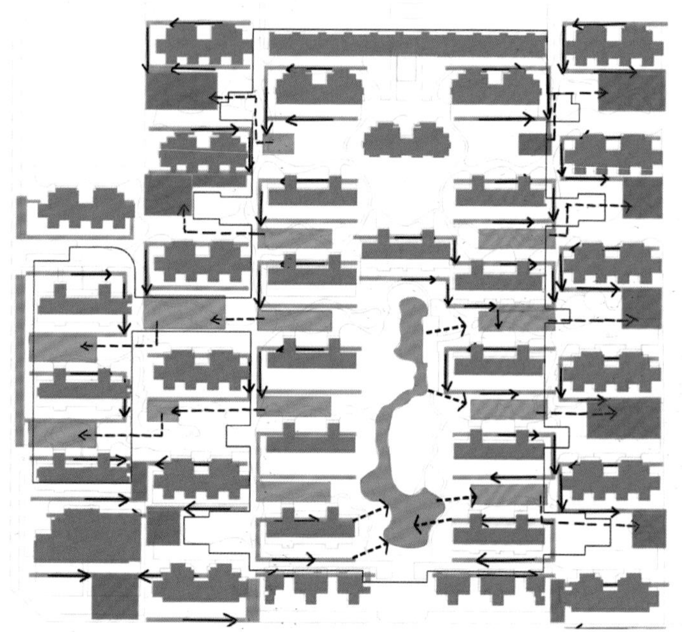

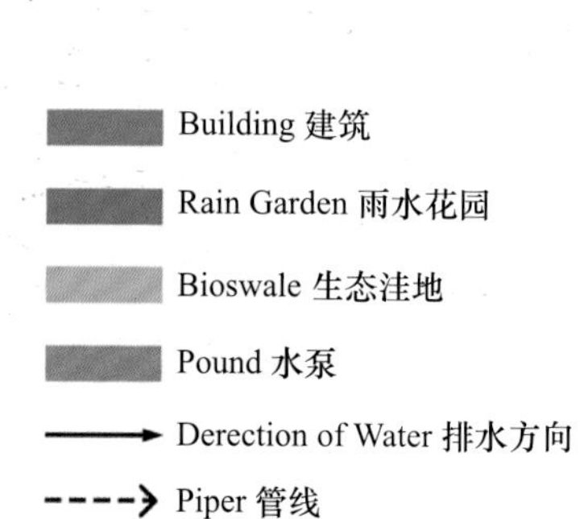

图 54 Strategic plan
规划平面

Calculation of our design 设计计算

A. change possible builidings roof to green roof（Max=12940.86m^2）with detensive plants

尽可能将普通屋顶改为绿色屋顶

B. expand the green land to the extrem (retain at least 15% of the ground for road)

扩展绿地（预留 15% 地面作为道路）

C. change pavement to the permeable one

将路面改为透水路面

D. add the possible detension pond（Max≈6550m^2）

加上可能的滞留池

E. add the possible rain garden（Max≈6960m^2）

加上可能的雨水花园

F. add the possible bioswale（Max≈6960m^2）

添加可能的生态洼地

Plan A: 50%A+15%B+15%C+50%D+0%E+50%F (According to UBC Calculator)
方案 A:50%A+15%B+15%C+50%D+0%E+50%F（使用 UBC 计算器）

表 24 Results of plan A
方案 A 计算结果

Reduced Final Runoff:
减少最终径流量
9251.24m^3

Require:
6532.24m^3
for green roofs irrigation
生态屋顶所需灌溉量

7 tanks (V=3000m^3)
7 个水箱

month 年份	X=A-B	A=Runoff from buildings 建筑径流	B=Roof Irrigation 屋顶所需灌溉	C=Remainder 存水量	Required irrigation 额外灌溉量	;0%D+0%E+50%F Result(m^2) 结果	Rate(%) 比例	Volume(m^3) 容量	Reduced runoff(m^3) 减少径流量	Strategy 策略
1	-376	0	376	0	376	6470.43	**50.00%**	/	According to the calculator 由计算器得出	A
2	-602	3	605	0	605					
3	-875	121	996	0	875					
4	-1832	0	1832	0	1832	19600.87	/	/		
5	-2402	11	2413	0	2402	29011.20	**15.00%**	/	According to calculator 由计算器得出	B
6	-1477	377	1854	0	1477					
7	1099	1515	416	1099	0	4170.00	**50.00%**	2085		D
8	1715	1715	0	1715	0					
9	-1656	18	1674	1158	0	2430.00	**50.00%**	1215		F
10	900	900	0	**2058**	0	24509.38	/	/		
11	-618	0	618	1440	0	27389.46	**15.00%**	/	According to calculator 由计算器得出	C
12	-408	0	408	1032	0					
1	-376	0	376	656	0					
2	-602	3	605	54	0					
3	-875	121	996	0	821	4108.42	**15.00%**	/	**987.20**	C
4	-1832	0	1832	0	1832					
5	-2402	11	2413	0	2402					
6	-1477	377	1854	0	1477	0.00	**0.00%**	0	According to calculator 由算计器得出	E
7	1099	1515	416	1099	0	1980.00	**50.00%**	990		F
8	1715	**1715**	0	1715	0	18969.52	/			
9	-1656	18	1674	1158	0	6470.43				
10	900	900	0	2058	0	26939.38				
11	-618	0	618	1440	0	52782.07				
12	-408	0	408	1032	0	20949.52				
GREY ROOF (m^2) 硬质屋顶						27389.46				

Runoff from Buildings(m^3) 建筑径流	4661.57	Roof Irrigation(m^3) 屋顶灌溉	11568	extra required irrigation(m^3) 所需额外灌溉	6532
Number of Tanks (V=10m*10m*3m) 水箱数量	7	Storage of Tanks(m^3) 水箱容量	**2100**	Detention Pond Storage(m^3) 水池容量	2085
Roof Runoff in UBC calculator(m^3)UBC 计算出的屋顶径流	14476.32	Garage Runoff(m^3) 停车场径流	7729.75	Final Runoff 最终径流 **2758.6**	Reduction(%) 减少率 **77.03%**

Plan B: 100%A+0%B+15%C+50%D+50%E+50%F (According to UBC Calculator)
方案 B:100%A+0%B+15%C+50%D+50%E+50%F（使用 UBC 计算器）

表 25 Results of plan B
方案 B 计算结果

Reduced Final Runoff:
减少最终径流量
7865.54m³

Require:
6536m³
for green roofs irrigation
生态屋顶所需灌溉量

9 tanks (V=3000m³)
9 个水箱

month 年份	X=A-B	A=Runoff from buildings 建筑径流	B=Roof Irrigation 屋顶所需灌溉	C=Remainder 存水量	Required irrigation 额外灌溉量	;0%D+0%E+50%F Result(m²) 结果	Rate(%) 比例	Volume(m³) 容量	Reduced runoff(m³) 减少径流量	Strategy 策略
1	-367	0	367	0	376	12940.86	**100.00%**	/	According to the calculator 由计算器得出	A
2	-601	2	603	0	601					
3	-960	110	970	0	960					
4	-1828	0	1828	0	1828	13130.44	/	/		
5	-2401	4	2405	0	2401	32320.24	**0.00%**	/	According to calculator 由计算器得出	B
6	-1453	345	1798	0	1453					
7	1124	1489	365	1124	0	4170.00	**50.00%**	2085		D
8	1646	1646	0	**2770**	0					
9	-1659	7	1666	1111	0	2430.00	**50.00%**	1215		F
10	886	886	0	1997	0	21200.34	/	/		
11	-616	0	616	1381	0	30943.33	**0.00%**	/	According to calculator 由计算器得出	B
12	-407	0	407	974	0					
1	-367	0	367	607	0					
2	-601	2	603	6	0	4641.50	**15.00%**	/	**1115.30**	C
3	-860	119	979	0	854					
4	-1828	0	1828	0	1828					
5	-2401	4	2405	0	2401	3430.00	**50.00%**	1715	According to calculator 由计算器得出	E
6	-1453	345	1798	0	1453	1980.00	**50.00%**	990		F
7	1124	1489	365	1124	0	11985.65	/			
8	1646	1646	0	2770	0	12940.86				
9	-1659	7	1666	1111	0	23630.34				
10	986	986	0	1997	0	49620.68				
11	-616	0	616	1381	0	17395.65				
12	-407	0	407	974	0	30943.33				

Runoff from Buildings(m³) 建筑径流	4499.12	Roof Irrigation(m³) 屋顶灌溉		11034	extra required irrigation(m³) 所需额外灌溉		**6536**
Number of Tanks (V=10m*10m*3m) 水箱数量	**9**	Storage of Tanks(m³) 水箱容量		**2700**	Detention Pond Storage(m³) 水池容量		2085
Roof Runoff in UBC calculator(m³)UBC 计算出的屋顶径流	14479.38	Garage Runoff(m³) 停车场径流		7965.26	Final Runoff 最终径流	**4144.3**	Reduction(%) 减少率 **65.49%**

Plan C: 0%A+15%B+15%C+50%D+50%E+50%F (According to UBC Calculator)
方案 C:0%A+15%B+15%C+50%D+50%E+50%F（使用 UBC 计算器）

表 26　Results of plan C
方案 C 计算结果

Reduced Final Runoff:
减少最终径流量
9406.64m³

Require:
5673m³
for green roofs irrigation
生态屋顶所需灌溉量

10 tanks (V=3000m³)
10 个水箱

month 年份	X=A-B	A=Runoff from buildings 建筑径流	B=Roof Irrigation 屋顶所需灌溉	C=Remainder 存水量	Required irrigation 额外灌溉量	;0%D+0%E+50%F Result(m^2) 结果	Rate(%) 比例	Volume(m^3) 容量	Reduced runoff(m^3) 减少径流量	Strategy 策略
1	-328	0	328	0	328	0.00	**0.00%**	/	According to the calculator 由计算器得出	A
2	-536	5	541	0	536					
3	-770	124	903	0	779					
4	-1636	0	1636	0	1636	26071.30	/	/		
5	-2134	23	2157	0	2401	29011.20	**15.00%**	/	According to calculator 由计算器得出	B
6	-1285	416	1701	0	1453					
7	1141	1557	416	1141	0	4170.00	**50.00%**	2085		D
8	1788	1788	0	**2929**	0					
9	-1467	30	1497	1462	0	2430.00	**50.00%**	1215		F
10	914	914	0	2376	0	24509.38	/	/		
11	-552	0	552	1824	0	27389.46	**15.00%**	/	According to calculator 由计算器得出	B
12	-364	0	364	1460	0					
1	-328	0	328	1132	0					
2	-536	5	541	596	0	4108.42	**15.00%**	/	**987.20**	C
3	-779	124	903	0	183					
4	-1636	0	1636	0	1636					
5	-2134	23	2157	0	2401	3430.00	**50.00%**	1715	According to calculator 由计算器得出	E
6	-1285	416	1701	0	1453	1980.00	**50.00%**	990		F
7	1141	1557	416	1141	0	15539.52	/			
8	1788	1788	0	2929	0	0.00				
9	-1467	30	1497	1462	0	26939.38				
10	914	914	0	2376	0	59252.50				
11	-552	0	552	1824	0	20949.52				
12	-364	0	364	1460	0	27389.46				

Runoff from Buildings(m^3) 建筑径流	4855.64	Roof Irrigation(m^3) 屋顶灌溉		10095	extra required irrigation(m^3) 所需额外灌溉		**5673**
Number of Tanks (V=10m*10m*3m) 水箱数量	**10**	Storage of Tanks(m^3) 水箱容量		**3000**	Detention Pond Storage(m^3) 水池容量		2085
Roof Runoff in UBC calculator(m^3)UBC 计算出的屋顶经流	14619.9	Garage Runoff(m^3) 停车场径流		7678.26	Final Runoff 最终径流	**2603.2** Reduction(%) 减少率	**78.32%**

3.2 Design Result of group two 第二组设计成果

3.2.1 Design ideas 设计思路

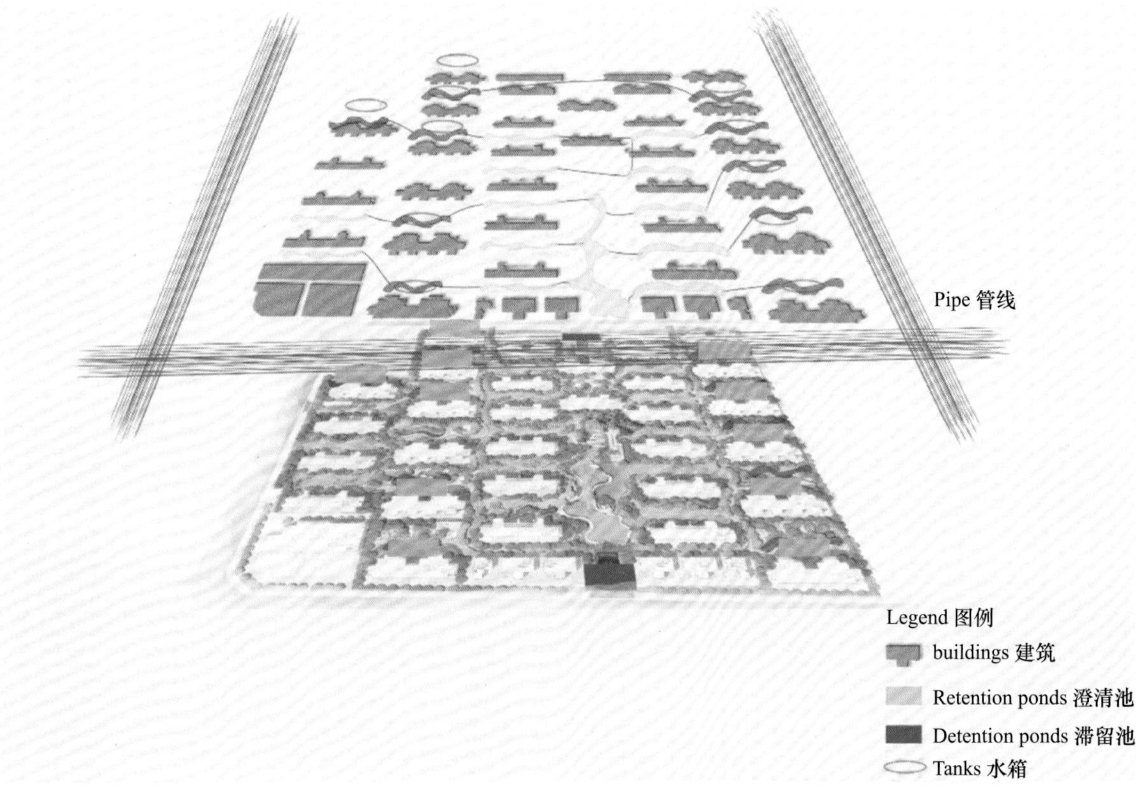

图 55 Site plan
场地平面图

· Areasonable proportion of greening plant configuration in residential areas is Trees: Shrubs: Grass=1:6:20.
居住区绿化植物配置的合理比例是：乔木：灌木：草木 =1 ： 6 ： 20。
So we set the data of green roof types to types to extensive roof: intensive roof =20:7
所以我们将生态屋顶类型的数据设置为粗放型：集约型 =20 ： 7。

· Passive housing and sloped roof can be designed as extensive roof.
被动房和坡屋顶可以被改造为粗放型生态屋顶。

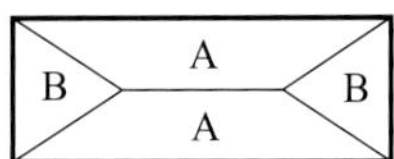

Actuall, green roof is just suit for A parts, like this.
事实上，生态屋顶只适用于 A 部分，如左图所示

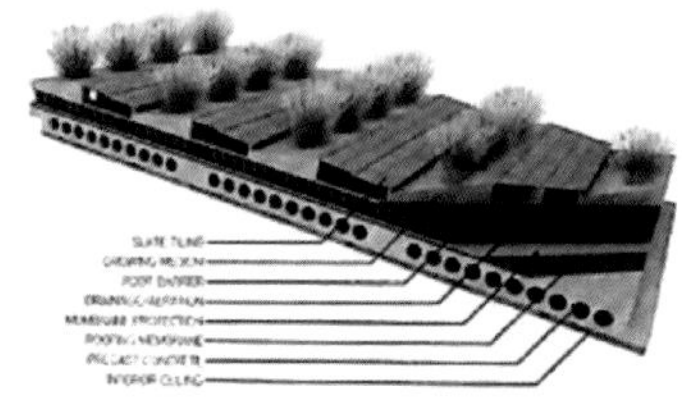

· Permeable paving: impermeable paving=1 ： 1
透水铺装：不透水铺装 =1 ： 1

· Grey ground area=15% Ground area * 50%
铺装面积 =15% 地面面积 × 50%

图 56 Design Assumptions
设计假设

表 27　Site surface classification and calculation coefficient
场地表面分类与计算系数

Roof 屋顶	面积	Roof Area 屋顶面积	Extensive Rf 粗放型生态屋顶	Intensive Rf 集约型生态屋顶	Grey Roof 硬质屋顶
	Area	97671.94	17450	9507.51	50614.33
	CN	—	—	—	98
	Kc	—	0.3	0.6	—

Ground 地面	面积	Ground Area 地面面积	Green 绿地	Grey 硬质铺装
	Area	46144.93	42693.97	3450.65
	CN	—	61	98

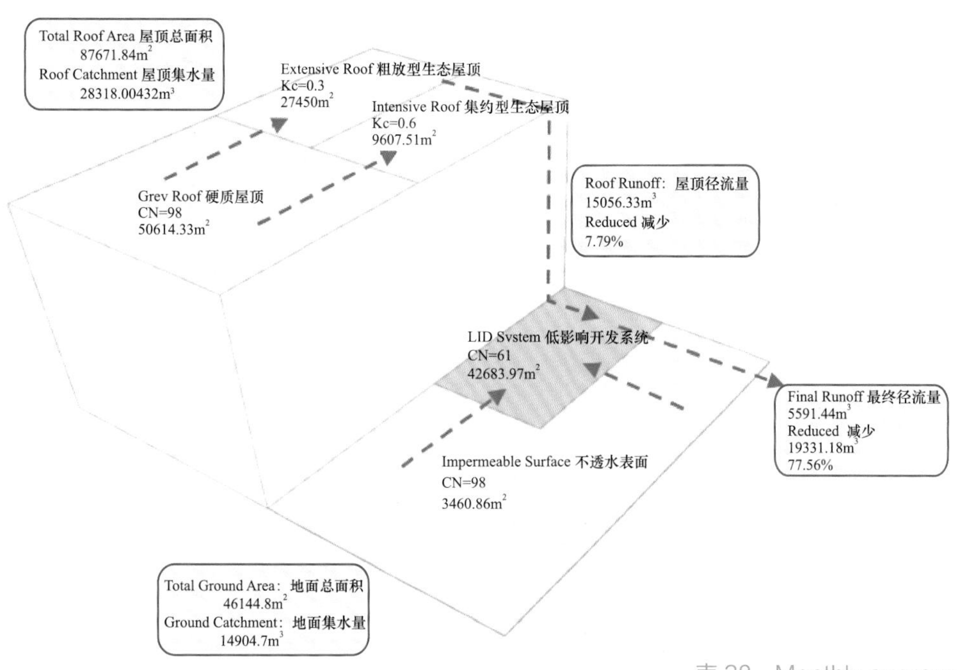

图 57　Data in LID calculator
LID 计算

表 28　Monthly average runoff calculation of green roof
生态屋顶月均径流量计算

Month 月份	Precipitation 降水	Roof Evapet ranspertation 屋顶蒸发	Roof Runoff 屋顶径流	Roof Irrigation 屋顶灌溉	Green Absorb 绿地吸收	All Runoff 总径流量
1	0	284	0	284	0	0
2	321	473	6	466	15	0
3	1298	893	400	717	619	0
4	0	1417	0	1417	0	0
5	923	1889	11	1856	52	0
6	4991	1845	1146	1255	1805	76
7	11883	1753	4995	148	4674	3077
8	15683	1575	5504	0	8084	583
9	1151	1324	21	1281	74	0
10	6972	716	2973	0	2736	1855
11	0	478	0	478	0	0
12	0	315	0	315	0	0

Precipitation and Water Mitigation Estimates 降水与洪水缓和评价
☑ Precipitation
☑ Roof_Evapetranspertation　☑ Roof_Runoff　☑ Roof_Irrigation
☑ Green_Infiltration　☑ Final_Runoff
Unit:m3
18000
16000
14000
12000
10000
8000
6000
4000
2000
0
月份 Month　1 Jan　2 Feb　3 Mar　4 Apr　5 May　6 June　7 July　8 Aug　9 Sep　10 Oct　11 Nov　12 Dec

图 58　Calculation of tank
水箱计算

表 29　Irrigation of roof
屋顶灌溉

Month 月份	Extensive roof(m^3) 粗放型生态屋顶	Intensive roof(m^3) 集约型生态屋顶	All (m^3) 总计
1	167.16	117.01	284.17
2	273.18	192.97	466.15
3	395.19	322.16	717.35
4	833.39	583.37	1416.76
5	1086.83	769.30	1856.13
6	647.97	606.69	1254.66
7	0.00	148.20	148.20
8	0.00	0.00	0.00
9	747.04	534.04	1281.08
10	0.00	0.00	0.00
11	281.00	196.70	477.70
12	185.51	129.86	315.36
All	4617.26	3600.30	8217.56

3.2.2 Holistic design 整体设计

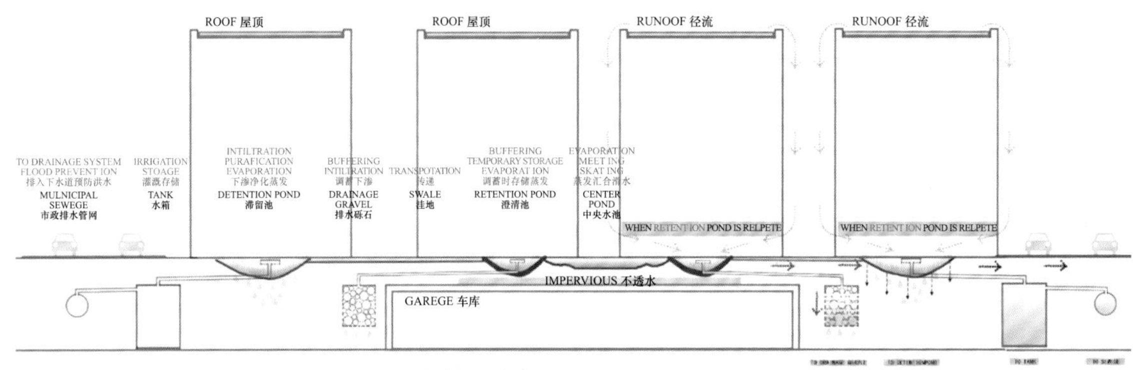

图 59 Holistic water system
整体水系统

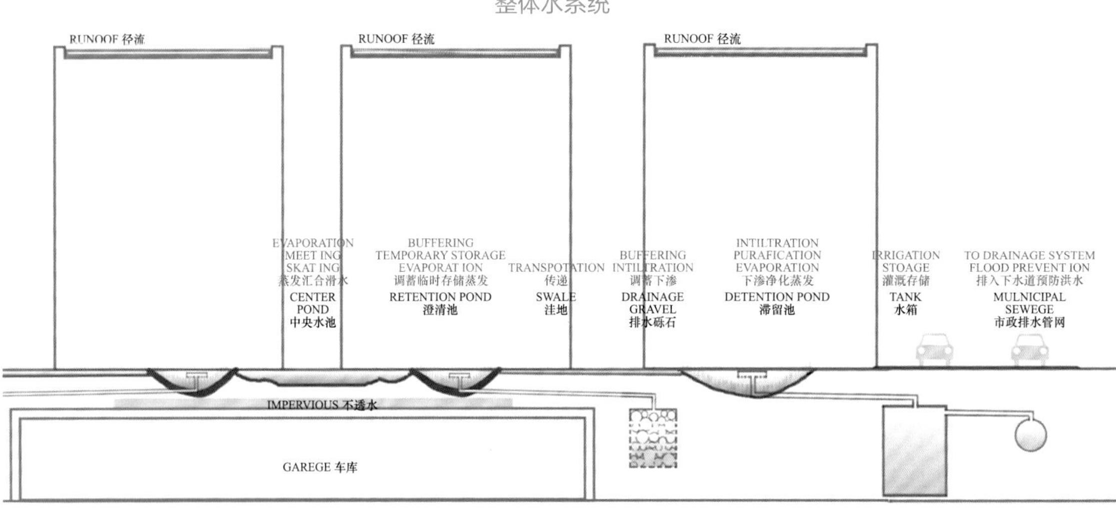

图 60 Component
系统构成

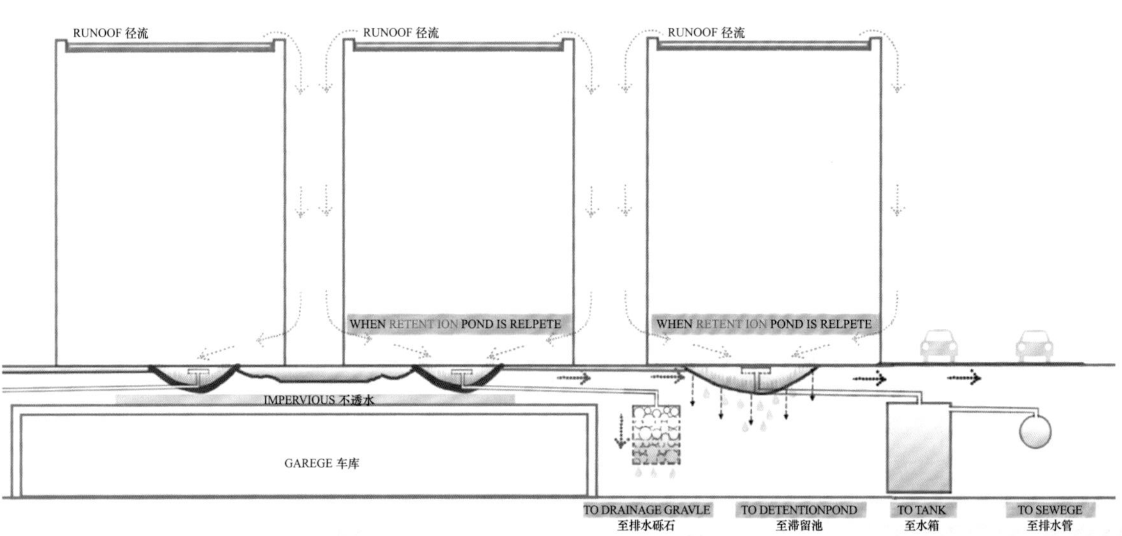

图 61 Rain path
雨水路径

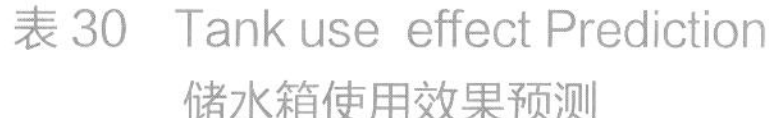

表 30　Tank use effect Prediction
储水箱使用效果预测

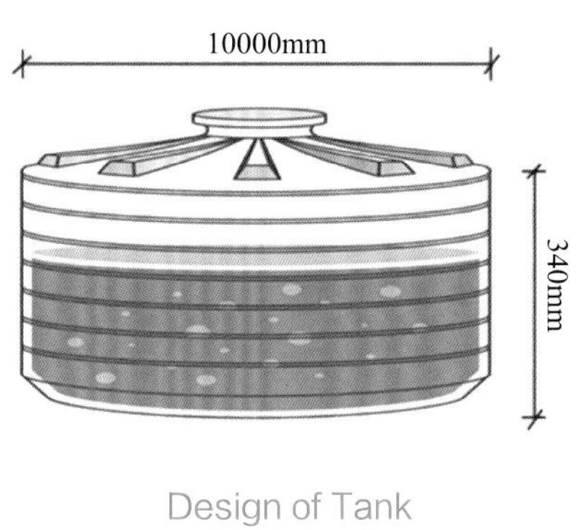

Design of Tank
储水箱设计

Month 月份	runoff 径流	irrigation 灌溉	water in tank 水箱中的水	buy extra water 额外需水
Jan	0	426	1763	
Feb	0	699	1064	
Mar	0	1076	0	12
Apr	0	2126	0	2126
May	0	2784	0	2784
Jun	76	1883	0	1807
Jul	3080	222	2858	
Aug	585	0	3443	
Sep	0	1922	1522	
Oct	1857	0	3379	
Nov	0	717	2662	
Dec	0	473	2189	
D=10m H=3.4m V=2π*D*H*13≈3469m³				

Surface/Plants　表面/植物
200~300mmSwale　200~300mm洼地
20~30mmOrganic mulch　20~30mm有机覆土
600~800mm Mixed soil　600~800mm混合土
Water Permeable Brick 透水砖
Sand-gravel cusion 砂砾石垫层
300~400mm Metalling　300~400mm喷镀金属
Plain soil 纯土
Plain soil　纯土
Sidewalks and bike paths
X行道与自行车道
1.5
1.5
Overflow pipe 溢流管
Drainpipe 排水管

图 62　Bio-swales section
生态洼地剖面

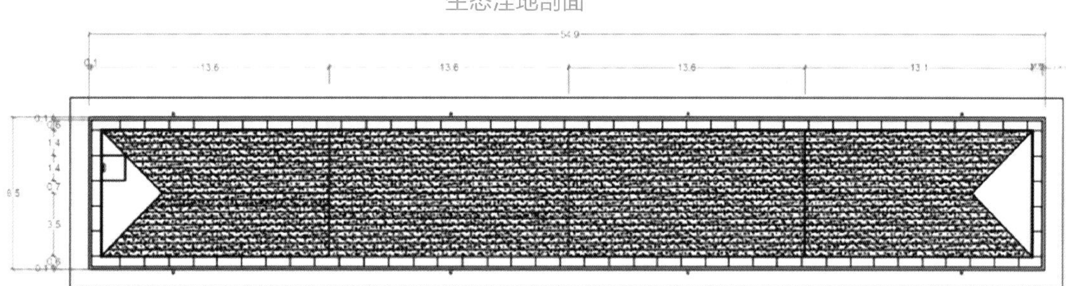

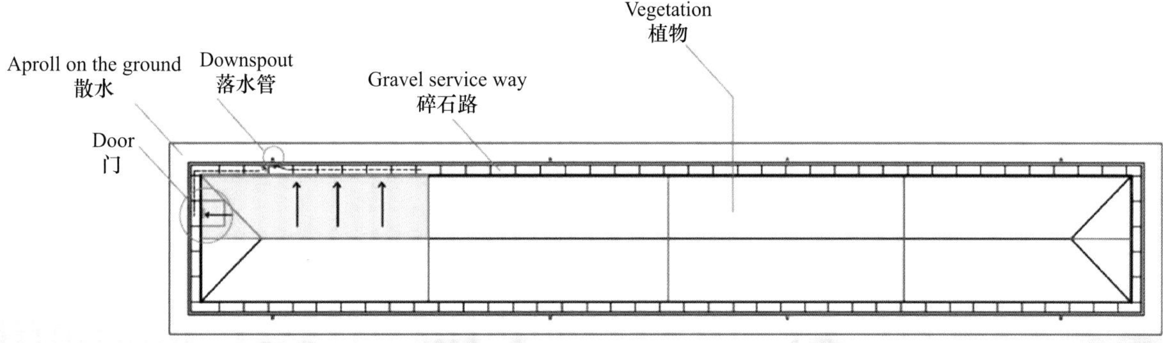

图 63　Extensive roof of buildings
建筑物屋顶设计

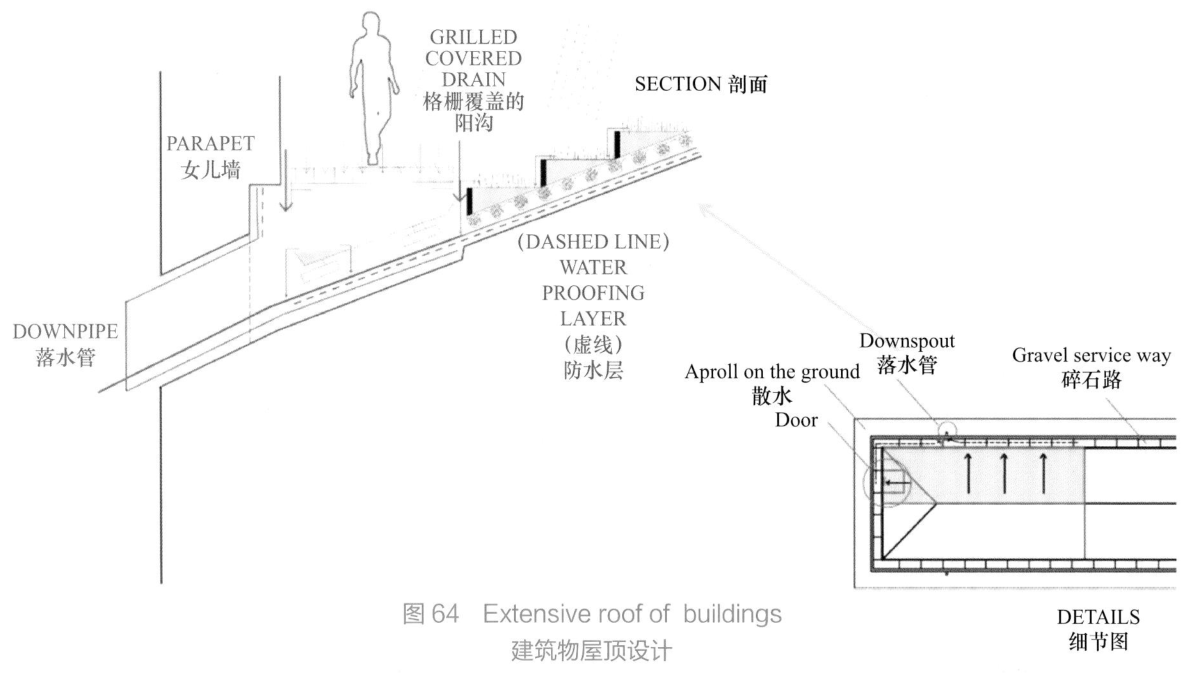

图 64　Extensive roof of buildings
建筑物屋顶设计

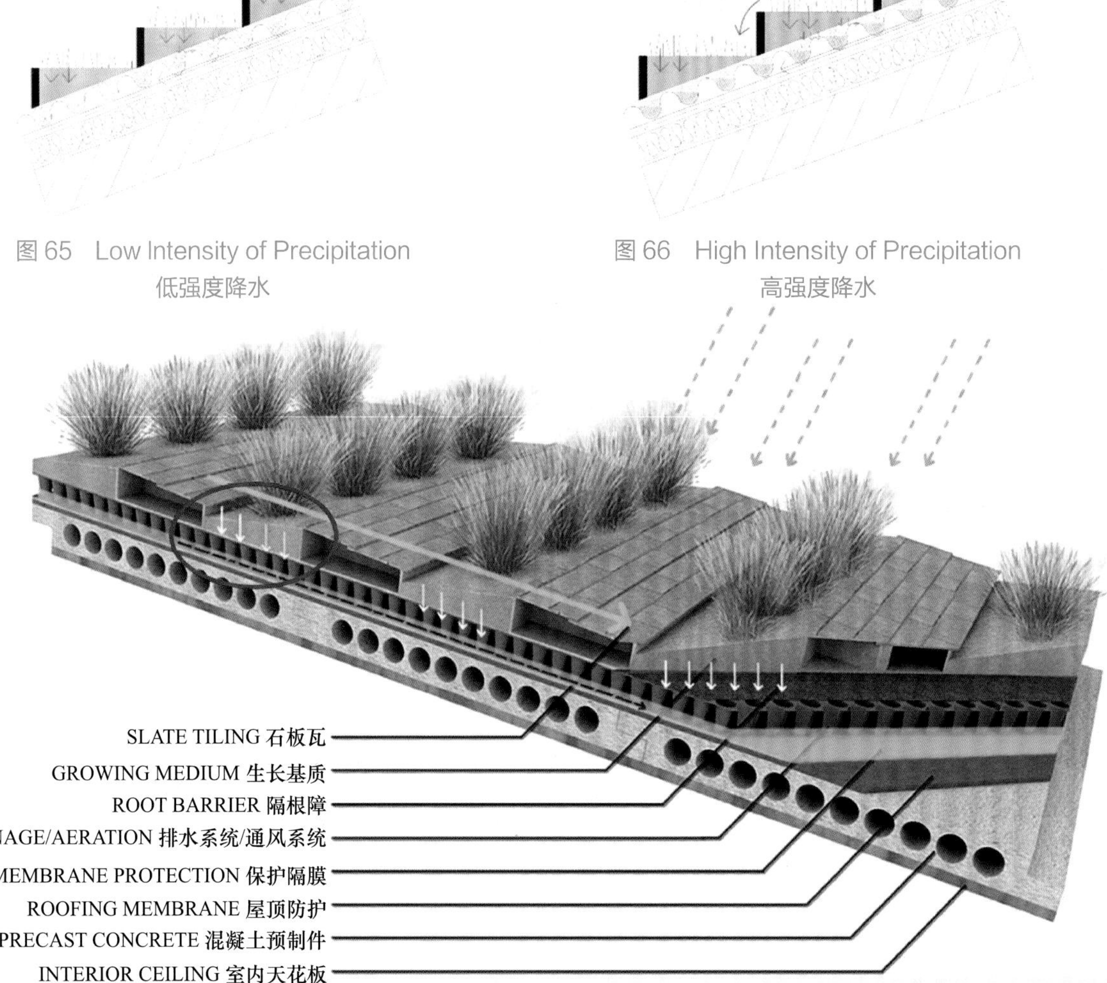

图 65　Low Intensity of Precipitation
低强度降水

图 66　High Intensity of Precipitation
高强度降水

图 67　Sloping green roof section
坡屋顶设计

3.2.3 Planting Plan 种植设计

	SPECIES 物种	NAME 名称	TYPE 类型	WATER CONSUMPTION 耗水量	HEIGHT 高度
	Miscanthus sinensis cv.	细叶芒	PERENNIAL 多年生		0.2~0.7m
	Imperata cylindrica (L) Beauv.	白茅	PERENNIAL 多年生		0.3~0.8m
	Pennisetum alopecuroides (L) Spreng.	狼尾草	PERENNIAL 多年生		0.3~1.2m
	Setaria viridis (L) Beauv.	狗尾草	ANNUAL 一年生		0.1~1m
	Poa annua L.	早熟禾	ANNUAL 一年生		0.07~0.3m
	Hemerocallis fulva	萱草 (Tiger Lily)	PERENNIAL 多年生		0.3~0.6m
	Zoysia japoncia	结缕草	PERENNIAL 多年生		0.03~0.1m

	SPECIES 物种	NAME 名称	TYPE 类型	WATER CONSUMPTION 耗水量	HEIGHT 高度
	Rosa chinensis Jacq.	月季	PERENNIAL 多年生		0.5~2m
	Malus micromalus	西府海棠	PERENNIAL 多年生		2~5m
	Forsythia suspensa	连翘	PERENNIAL 多年生		1~3m
	Philadelphus pekinensis Rupr.	山梅花	PERENNIAL 多年生		1.5~3.5m
	Amygdalus triloba	榆叶梅	ANNUAL 一年生		2~3m
	Ligustrum× vicaryi Hort	金叶女贞 (Hybrida Vicary Privet)	PERENNIAL 多年生		2~3m
	Buxus megistophylla Levl.	大叶黄杨	PERENNIAL 多年生		0.6~2.2m

	SPECIES 物种	NAME 名称	TYPE 类型	WATER CONSUMPTION 耗水量	HEIGHT 高度
	Robinia pseudoacacia L.	刺槐	PERENNIAL 多年生		10~25m
	Sophora japonica	国槐	PERENNIAL 多年生		15~25m
	Pinus thunbergii Parl.	黑松	PERENNIAL 多年生		20~30m
	Sophora japonica	龙爪槐	PERENNIAL 多年生		10~25m
	Populus Tomentosa Carr	白杨	PERENNIAL 多年生		20~30m
	Crape myrtus	紫薇	PERENNIAL 多年生		5~10m
	Magnolia Denudata Desr.	玉兰	PERENNIAL 多年生		10~25m

UBC+BFU “Green Community and Low Impact Development (LID) Frontier Design Workshop” Summer Course Ends Successfully

UBC+BFU“绿色社区与低影响开发 (LID) 前沿设计工作坊”暑期课程顺利结束

During this summer school (08.27-09.07), we were fortunate to have invited Professor Daniel Roehr from UBC and his assistant MUD Master Jericho Bankston to teach us about Low Impact Development (LID) with our Professor Li Yi and Lecturer Ge Xiaoyu.

本次小学期（08.27—09.07）我们有幸请到了来自 UBC 的 Daniel Roehr 教授和其助手 MUD 硕士 Jericho Bankston，同我校李翅教授和戈晓宇讲师一起为我们讲解有关低影响开发 (LID) 的相关内容。

Learning Contents:

The teachers made a total of twelve wonderful lectures involving:

Proposal of low impact development LID concept;

The importance of low impact design to deal with stormwater problems in the context of climate change;

Comparison of design standards and design strategies at home and abroad;

Principles, calculation logic, application conditions, results analysis and differences of stormwater calculation methods at home and abroad;

Specific measures and related links of Low Impact Development;

How to apply LID as a system for site design or transformation;

How to apply LID ideas for residential areas and urban planning;

Several aspects such as case study at home and

学习内容：

老师们总共做了十二个精彩的讲座，涉及：

低影响开发 LID 理念的提出；

在气候变化大背景下，低影响设计应对雨洪问题的重要性；

国内外设计标准及设计策略对比；

国内外雨洪计算方法的原理、计算逻辑、应用条件、结果分析及差异；

低影响开发具体措施以及相关联系；

如何应用 LID 作为一个体系进行场地设计或改造；

如何应用 LID 思想进行居住区、城市规划；

国内外案例学习等几个方面。

abroad.

In addition, we organized two site studies and researches.

除此之外，我们组织了两次场地学习和调研。

We conducted site research on the campus of Beijing Forestry University on August 27th. And we also made investigations and analysis on existing building water outlets, ground rainwater outlets, pavements, rainwater gardens and ecological grassland demonstration areas. Through discussions with teachers, we proposed preliminary campus renovation ideas.

8 月 27 日，我们在林大校园进行场地调研，针对现有建筑落水口、地面雨水口、地面铺装、雨水花园及生态植草沟示范区等进行调查和分析，通过和老师的讨论，提出初步校园改造思路。

We went to the National Green Smart Building Demonstration Center of Gaobeidian on August 29th.We visited the passive house combined with green roof, new permeable concrete floor, rainwater collecting tower, modular underground rainwater reservoir, rain garden, filter pool-plant ecology purification pool - biological ecological purification pool - clean water observation pool consists of four layers of ecological oxygen purification pool system.We learnt a lot from the design principles, engineering practices and other aspects. After that, we went to the constructing site of Longhu Train New City, and under the guidance of the teacher and the project team, conducting site research and conceived preliminary redesign ideas.

8 月 29 日，我们去到了高碑店国家绿色智慧建筑示范中心，参观了结合绿色屋顶的被动式房屋、新型透水混凝土地坪、雨水收集塔、模块式地下雨水蓄水池、雨水花园、由过滤池—植物生态净化池—生物生态净化池—净水观察池组成的四层生态曝氧净化池体系，从设计原理、工程做法等方面进行学习，收获颇丰。之后又去到了龙湖列车新城一区开发现场，在老师和方案设计团队的指导下，进行场地调研，构想初步再设计思路。

Learning experience:

After two weeks of study, we have a preliminary understanding and understanding of LID, learn to choose the appropriate LID measures according to the existing site conditions, integrate them into a system, and rationalize our design through runoff calculation and cost calculation.

学习感受：

通过两周的学习，我们对 LID 有了初步的认识和了解，学会了根据现有场地状况，选择合适的 LID 措施，将其整合为一个系统，通过径流计算、耗资计算让我们的设计合理化。

In the classroom, we deeply felt the attention of foreign teachers to the details, such as detailed cross-sections, elevations, plans, and structural practices, which were specifically shown in a complete design, water source, transmission, retention, collection, purification, utilization and other content. Through

课堂上，我们深刻感受到了国外老师对细节的重视，例如通过详细的剖面图、立面图、平面图、结构做法图，具体展示在一个完整的设计中，水的来源、传输、滞留、收集、净化、利用等内容。通过图示，我们明确了每个 LID 措施之间的关系，以及如何构建一个完整的 LID 系统。

the illustration, we clarify the relationship between each LID measure and how to build a complete LID system. In terms of calculation, we repeatedly refine each input and output data, and adjust the parameters to ensure that the calculation results are in line with the the objective facts.

In terms of presentation of results, we strive to visualize the results of the analysis. 'Make sure the graphics say what you want them to without having to explain.' The professional content is presented to the public in a clear and concise graphic language to help everyone build a low-impact development concept and promote low-impact development.

在计算方面，我们对每一个输入和输出的数据进行反复推敲，通过调节参数，保证计算结果和客观事实最大程度的相符。

在成果呈现方面，我们力求做到分析结果可视化。'Make sure the graphics say what you want them to without having to explain.'将专业的内容用简洁明了的图示化语言展示给公众，帮助每个人构建低影响开发的理念，推动低影响开发的进程。

Summary

During the 12-day green community and low-impact development (LID) frontier design workshops, everyone enjoys the process of group cooperation. Through continuous discussion, our inspirations collide with each other.We generate new ideas, and add highlights to the project. From the preliminary research and analysis, to the medium-term plan conception, to the later design deepening. We have conducted PPT report, exchanging and discussing in each stage in a timely manner. Although there were frequent problems in the design process, we can always be inspired and find reasonable answers and get a lot of new knowledge through group communication, data review, and discussion with teachers. This summer school has been successfully completed. Thanks to the teachers for their careful teaching. We have all gained a lot, and hope that the next summer school will be exciting as well.

总结

为期 12 天的绿色社区与低影响开发 (LID) 前沿设计工作坊，大家都很享受小组合作的过程，通过不断的讨论，让灵感相互碰撞，产生新的思路，为项目增添亮点。从前期调研分析，到中期方案构想，到后期设计深化，每个阶段我们都及时进行了 PPT 汇报交流与讨论。虽然整个设计过程中频频遇到问题，但通过组内交流、查阅资料、和老师们讨论，我们总能受到启发并找到合理的答案，学习到很多新的知识。本次小学期已经圆满结束了，感谢老师们的悉心教导，我们都收获颇丰，同时希望下一个小学期能一样精彩。

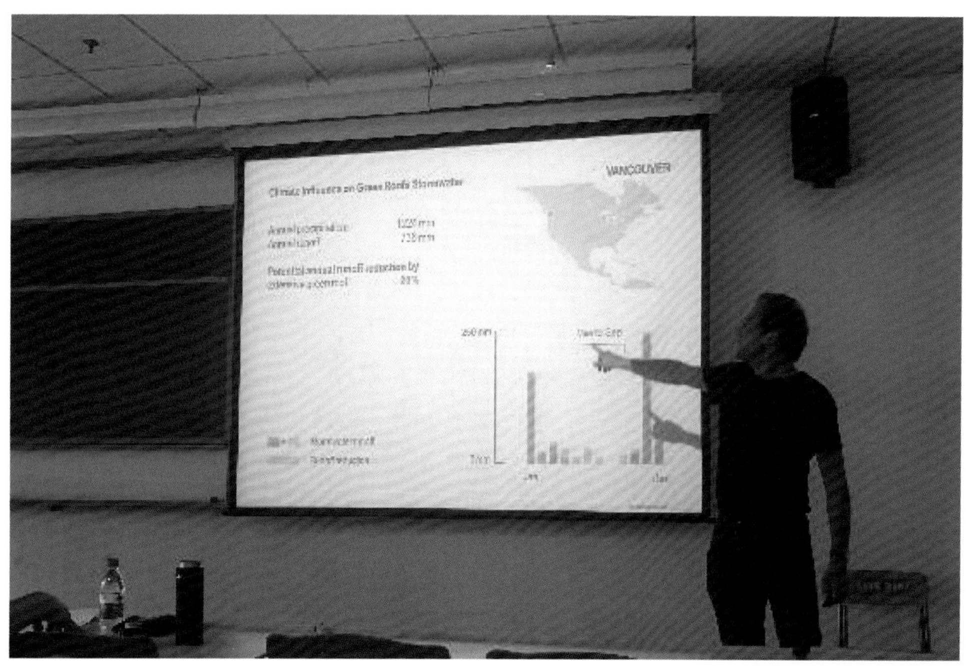

图 1　Professor Daniel Roehr is explaining climate characteristics of Vancouver
Daniel Roehr 教授在讲解温哥华地区气候特点

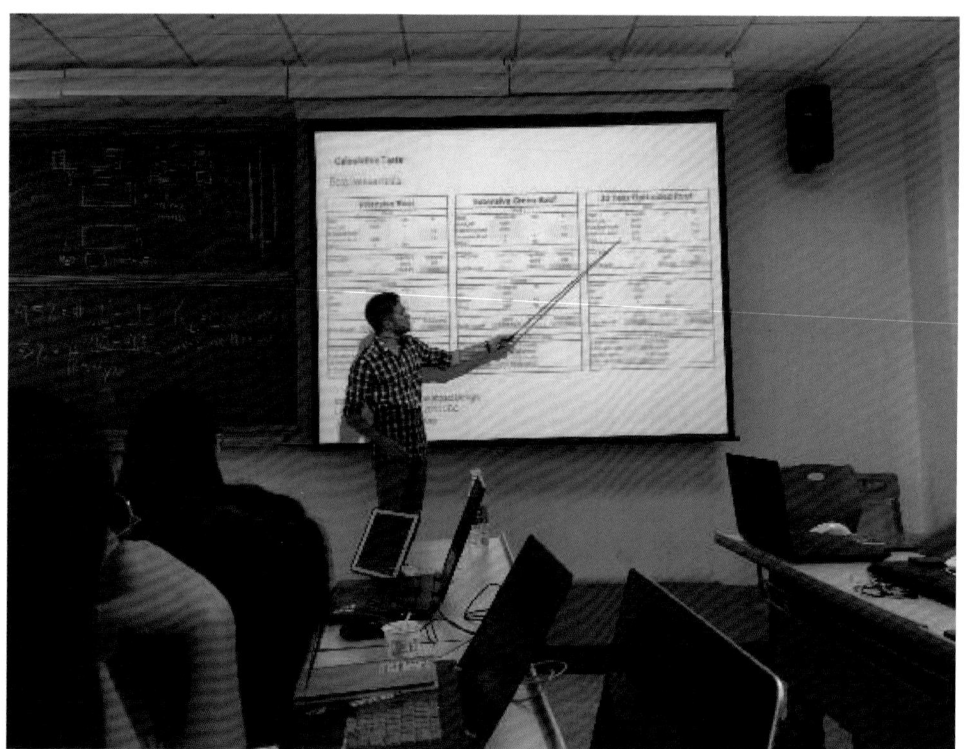

图 2　Jericho Bankston explaining the application of UBC calculator
Jericho Bankston 在讲解 UBC 计算器应用

图 3　Visiting the construction site of Rau New Town of Longhu
参观龙湖列车新城一区开发现场

图 4　Campus site survey
校园场地调研

图 5 Visit and study the new permeable concrete floor of Gaobeidian National Green Intelligent Building center
参观学习高碑店国家绿色智慧建筑示范中心的新型透水混凝土地坪

图 6 Visit and study the modular underground rainwater reservoir of Gaobeidian National Green Intelligent building center
参观学习高碑店国家绿色智慧建筑示范中心的模块式地下雨水蓄水池

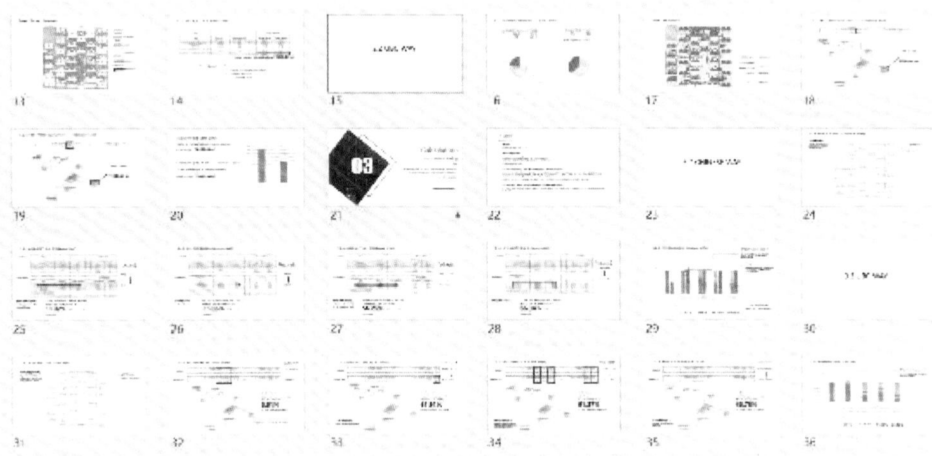

图 7　Achievements 1
成果展示 1

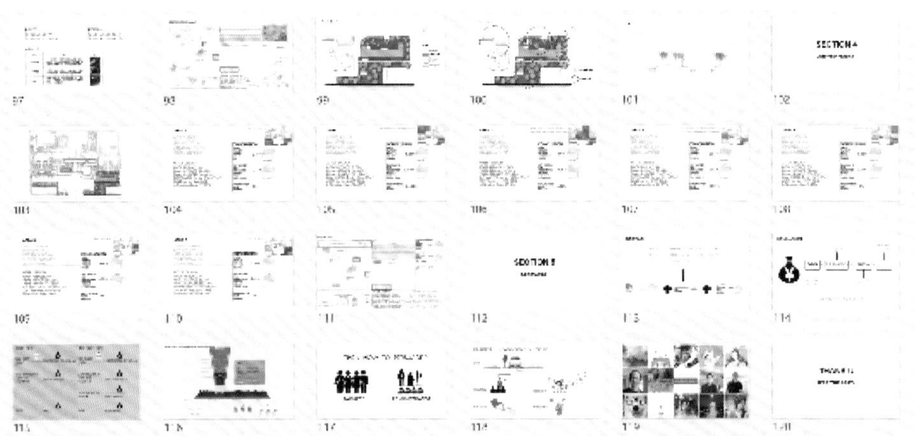

图 8　Achievements 2
成果展示 2

图 9　Group photo
全体合影